LLA
10340459
612.044 ROT

D0726419
61
RO

Primers in Exercise Science S

Genetics Primer for Exercise Science and Health

Stephen M. Roth, PhD
University of Maryland

UWIC LEARNING CENTRE
LIBRARY DIVISION-LLANDAFF
WESTERN AVENUE
CARDIFF
CF5 2YB

Human Kinetics

Library of Congress Cataloging-in-Publication Data

Roth, Stephen M., 1973-
Genetics primer for exercise science and health / Stephen M. Roth.
p. ; cm. -- (Primers for exercise science series)
Includes bibliographical references and index.
ISBN-13: 978-0-7360-6343-2 (soft cover)
ISBN-10: 0-7360-6343-9 (soft cover)
1. Human genetics. 2. Exercise--Physiological aspects. 3. Sports--Physiological aspects. I. Title. II. Series.
[DNLM: 1. Genetics. 2. Physical Fitness--physiology. 3. Exercise--physiology. 4. Genetic Techniques. 5. Variation (Genetics) QU 500 R845g 2007]
QH431.R84 2007
612'.044--dc22
2007004421

ISBN-10: 0-7360-6343-9
ISBN-13: 978-0-7360-6343-2

Copyright © 2007 by Stephen M. Roth

All rights reserved. Except for use in a review, the reproduction or utilization of this work in any form or by any electronic, mechanical, or other means, now known or hereafter invented, including xerography, photocopying, and recording, and in any information storage and retrieval system, is forbidden without the written permission of the publisher.

The Web addresses cited in this text were current as of March 15, 2007, unless otherwise noted.

Acquisitions Editor: Michael S. Bahrke, PhD; **Developmental Editor:** Christine M. Drews; **Assistant Editor:** Maureen Eckstein and Laura Koritz; **Copyeditor:** Joyce Sexton; **Proofreader:** Sarah Wiseman; **Indexer:** Craig Brown; **Permissions Manager:** Carly Breeding; **Graphic Designer:** Fred Starbird; **Graphic Artist:** Denise Lowry; **Photo Asset Manager:** Laura Fitch; **Photo Office Assistant:** Jason Allen; **Cover Designer:** Robert Reuther; **Art Manager:** Kelly Hendren; **Illustrator:** Argosy; **Printer:** United Graphics

Printed in the United States of America 10 9 8 7 6 5 4 3 2 1

Human Kinetics
Web site: www.HumanKinetics.com

United States: Human Kinetics, P.O. Box 5076, Champaign, IL 61825-5076
800-747-4457
e-mail: humank@hkusa.com

Canada: Human Kinetics, 475 Devonshire Road Unit 100, Windsor, ON N8Y 2L5
800-465-7301 (in Canada only)
e-mail: orders@hkcanada.com

Europe: Human Kinetics, 107 Bradford Road, Stanningley, Leeds LS28 6AT, United Kingdom
+44 (0) 113 255 5665
e-mail: hk@hkeurope.com

Australia: Human Kinetics, 57A Price Avenue, Lower Mitcham, South Australia 5062
08 8372 0999
e-mail: liaw@hkaustralia.com

New Zealand: Human Kinetics, Division of Sports Distributors NZ Ltd., P.O. Box 300 226 Albany, North Shore City, Auckland
0064 9 448 1207
e-mail: info@humankinetics.co.nz

To my wife, Nancy, and our children, Nels, Ellyn, and Anna

Contents

UWIC LEARNING CENTRE
LIBRARY DIVISION-LLANDAFF
WESTERN AVENUE
CARDIFF
CF5 2YB

Preface

Any familiarity with the news media has introduced you to the idea that genetics has an important place in modern health and medicine. The Human Genome Project was a major news event in the early part of the 21st century, with its identification of the more than 3 billion "letters" of the human genetic sequence. Perhaps more importantly, we all have an intuitive sense of the importance of genetics: We often resemble our parents and other family members; the idea of "good genes" influencing mental or physical performance is common throughout modern societies; and the focus on genetics as a tool in the fight against diseases such as cancer and obesity is gaining strength in public debate. As a student or professional in the areas of health, physical activity, or sport, no doubt you've wondered about a place for genetics within your sphere of interest or expertise as well. Certainly the idea that genetic variation can influence disease susceptibility and sport performance is common, and even the concept of "gene doping" for enhanced sport performance has gained notice in the media. Less well appreciated is the idea that genetic variation may have potential importance in the area of diet or exercise prescription for treating or preventing common diseases, though several research groups are focused on these areas.

This book aims to introduce students and professionals to the basic concepts of genetics as they relate to human health, physical activity, and sport. There is little in the way of an entry-level textbook to provide introductory concepts in genetics as it relates to the umbrella fields of exercise science and health. This book seeks to fill that void by providing the basic information needed to understand the role of genetics in these areas, interpret research findings, and begin to incorporate genetics into research programs. The text will also provide a structure for professionals interested in developing course work for upper-level undergraduates and graduate students. As the title implies, this is not a comprehensive analysis of DNA biology or genetics, but rather a text designed to provide the information necessary to interpret research findings and understand the anticipated importance of genetics in the future practice of exercise science and health.

The text is organized into three major parts that follow the logic of incorporating specific skills and concepts in a progressive fashion culminating in a strong breadth of information with modest depth. In part I, the basics of DNA and genetics are presented to give readers the information needed to apply this knowledge to the contexts of health, physical activity, and sport. While chapters 1 through 6 provide readers with skills to interpret much of the existing literature (with help from part II), future research will become more complex, and both students and professionals will need more advanced skills to interpret these more sophisticated studies. Thus chapter 7 provides more advanced information on DNA structure and genetics. Perhaps more importantly, the advanced genetics topics that are discussed at the end of part I offer a more complete picture of the complexity of genetic variation and its influence on physical traits and health.

With this foundation laid, part II describes specific skills and strategies for applying and interpreting genetics in research, including strategies for identifying research literature, understanding study design, and interpreting research findings. I finish this section with an overview of the typical laboratory methods used in the study of genetics. Part III of the text takes a step back from the details and presents a broader overview of the use of genetics in exercise science and health. Chapter 11 describes a few concrete examples of the effects of genetic variation on health- and physical activity-related variables. To provide an update on current research findings in a field moving as fast as genetics would be impossible, so the intent of this chapter is to give concrete examples in order for readers to gain experience in the critique and interpretation of typical research results. Readers are also introduced to the likely application of genetics to "clinical practice" in chapter 12, be it in the medical clinic or the fitness facility. Finally, with the continued advancement and use of genetic technologies in the health and fitness fields comes the potential for misuse of those technologies. A discussion of the ethical issues raised by genetics in society is presented in the final chapter. Throughout the text, readers will find "Special Focus" sections, which are expanded discussions of interesting topics related to a particular chapter.

Despite the complexities inherent in human genetics, the fact remains that individual genetic differences do influence health and performance measures, which means that genetics will forever have a place in the areas of health, physical activity, and sport. As an educator with background in both exercise science and human genetics, my primary goal with this text is to provide students and professionals with the basic information needed in order to feel comfortable in this emerging area. I expect that my attempts to present the necessary knowledge and skills in a clear and informative way may fall short in a few places, and so I encourage your feedback to improve any future editions of the book.

This book is the second volume in Human Kinetics' *Primers in Exercise Science Series,* which provides students and professionals alike with a nonintimidating basic understanding of the science behind each topic in the series, and where appropriate, of how that science is applied. These books, written by leading researchers and teachers in their respective areas of expertise, present in an easy-to-understand manner essential concepts in dynamic, complex areas of scientific knowledge. The books in the series are ideal for researchers and professionals who need to obtain background in an unfamiliar scientific area or as an accessible basic reference for those who will be returning to the material often.

This particular book adds a health dimension to the exercise science focus of the series. In addition to covering aspects of genetics related to physical activity, fitness, and sport performance, a good portion of the book discusses disease- and health-related information. Genetics has a long history of research related to disease, especially regarding what genes influence disease or cause medical problems, and a much shorter history related to physical activity and sport performance. Even where physical activity has been considered, it has typically been with an eye toward how physical activity might affect the function of a gene in order to prevent disease. Of course, recent endeavors to use genetics to improve sport performance take this research in a different direction. This book should provide exercise scientists and health professionals with a basic understanding of genetics as it relates to their fields.

Acknowledgments

I would not be in the field of exercise science were it not for the fact that Dr. Brent Ruby at the University of Montana opened his lab to an eager undergraduate in search of a start in research several years ago. My work then with Drs. Brent Ruby and Brian Sharkey stimulated an interest in physiology that has only grown with time. As a graduate student, I was fortunate to have the mentorship of Drs. Marc Rogers, Ben Hurley, and Jim Hagberg at the University of Maryland, the last of whom stimulated my initial interest in genetics. My expertise in genetics comes in large part from my time as a postdoctoral fellow under Dr. Robert Ferrell at the University of Pittsburgh. Finally, the students and fellows at the University of Maryland are important sounding boards for my ideas, and they ask the tough questions that keep me motivated to continue moving forward. Thank you all.

The decision to take on the task of writing this book was not easy, and I appreciate the hard work and support of Dr. Mike Bahrke, Christine Drews, Maureen Eckstein, and other staff at Human Kinetics throughout the process.

Finally, I am continually thankful to my wife, Nancy, and our children, Nels, Ellyn, and Anna, for their continued support in my career endeavors.

PART

I

BASICS OF GENETICS

In part I, we begin to address the most basic questions about DNA and genetics: What is DNA and how does it carry information for our cells? How does the cell use that information to perform necessary cellular activities? How are DNA and thus genetic information passed along to the next generation? How does your unique complement of DNA affect your health or performance? Before we can address these questions, however, we need to justify the study of genetics in the first place. Part I provides the basic information on DNA, cell and molecular biology, and genetics needed to begin understanding and applying genetics within the contexts of health, physical activity, and sport. Chapter 1 begins by examining the basic idea that genetic information is important to many of the physical traits and measurements that health and fitness professionals study. Chapter 2 focuses on DNA, RNA, and proteins and their organization within the cell. Chapter 3 describes the genetic code and the movement of information from the DNA of the cell's nucleus ultimately to the proteins performing the work of the cell. Chapter 4 outlines the process of meiosis, the separation of DNA during the formation of sperm and egg that is critical to the movement of genetic information to subsequent generations. In chapter 5, we move to more pure genetics, with discussions of heritability and genetic variation. These topics transition into chapter 6, where we consider the link between genetic variation and disease risk. Finally, chapter 7 introduces readers to the concepts of linkage disequilibrium and haplotype, which are important for developing a more complete understanding of genetics in the contexts of exercise and health. This last chapter of part I is more complex and challenging material, but it reflects the true complexity of genetic influence. After reading chapter 7, readers are encouraged to move through the remaining parts of the book, and then revisit chapter 7 to more fully understand its contents and importance.

CHAPTER

1

INDIVIDUAL DIFFERENCES: THE ROLE OF GENETICS IN EXERCISE SCIENCE AND HEALTH

When one is seeking to understand human biology, a common focus is to seek to understand *typical* biology, or normal biological responses or adaptations to various stimuli. Knowing the typical adaptations to aerobic exercise training, fitness trainers and coaches can develop and "prescribe" exercise programs for improving aerobic fitness. Knowing typical responses of blood pressure to a certain medication, a physician can confidently write a prescription for a patient with hypertension. Thus, the underlying goal for researchers within the areas of health, physical activity, and even sport, is to understand the typical or average responses and adaptations to various acute and chronic therapies or interventions. This goal is well known and has formed the basis of research activities since the early beginnings of the sciences of health and physical activity. Even the study of disease abnormalities in medicine is used to advance understanding of typical biology and thus knowledge of how "normal" physiology becomes abnormal.

The labors of researchers in these diverse fields have thus identified such physiological characteristics as healthy levels of blood pressure and blood lipids, and their typical responses to various therapies; average or normative values of maximal oxygen consumption ($\dot{V}O_2max$) before and after aerobic exercise training; typical values of cognitive performance for healthy men and women at various ages; and average quadriceps muscle strength levels before and after strength training. In fact, finding such normative data for a wide range of physical characteristics is no longer a challenge, whether these data are published in research review articles or more broadly in textbooks or pronouncements from professional or government organizations.

HIDDEN ASPECT OF AVERAGE VALUES

Despite knowledge of typical responses and adaptations to many, many therapeutic interventions, there is an underlying error in the idea of average values. As the saying goes, no one is average! While the average values for any particular physical characteristic or **trait** are useful for *guessing* about the typical response or adaptation of that trait to a stimulus, very rarely are such predictions perfect for any one individual. Such normative data are formed from the analysis of many individuals and their unique responses and adaptations, which when combined together form the **average** values used as the starting points for intervention, prescription, and so on.

Let's consider an example from the area of physical activity. While exercise scientists have been broadly successful in defining responses and adaptations to various exercise stimuli, a key feature of these responses and adaptations is that they represent *average* or normative values. In other words, responses

KEY POINT

While useful as indicators of typical values of a physical trait, average values hide the extent of variability found in the individual values that make up that average.

and adaptations to exercise activities differ remarkably among individuals. As an example, figure 1.1 shows the change in maximal oxygen consumption ($\dot{V}O_2max$) for individuals who participated in a large, multicenter exercise training trial, known as the HERITAGE Family Study. Maximal oxygen consumption is assessed by a maximal exercise test, and it represents the maximal capacity for oxygen consumption by the body, including the transport of oxygen from the lungs to the working muscles; aerobic exercise training is an excellent means of improving maximal oxygen consumption, which is sometimes referred to as aerobic fitness. Subjects in this research study performed 20 weeks of aerobic exercise training designed to improve cardiovascular and metabolic risk factors, as well as improve aerobic fitness. All subjects performed the same level of physical activity as measured by heart rate monitors (i.e., performed exercise at the same percentage of baseline $\dot{V}O_2max$), which was adjusted throughout the intervention to account for improvements in fitness. At both the beginning and end of the study period, each participant underwent a standard maximal exercise test to determine $\dot{V}O_2max$, which is expected to increase in response to aerobic exercise training.

Figure 1.1 shows four graphs, one from each of the four research centers across North America where the HERITAGE Family Study was performed. Each graph shows the $\dot{V}O_2max$ change for each individual subject to the exercise training at each of the four centers (i.e., the Y-axis represents change in $\dot{V}O_2max$). Each individual is shown as a vertical bar on the X-axis, and individuals are ordered from lowest $\dot{V}O_2max$ change to highest. As can be seen at each of the four research centers, dramatic differences in the change in $\dot{V}O_2max$ in response to exercise training were found among the research participants. Notice that across all four graphs, the average change in $\dot{V}O_2max$ was ~400 ml/min, so that for any particular individual in the study, we would expect a 400 ml/min increase in $\dot{V}O_2max$ as a consequence of the training. To the contrary, at all four centers we see several individuals with little to no improvement in $\dot{V}O_2max$ despite 20 weeks of aerobic exercise training! Such a response

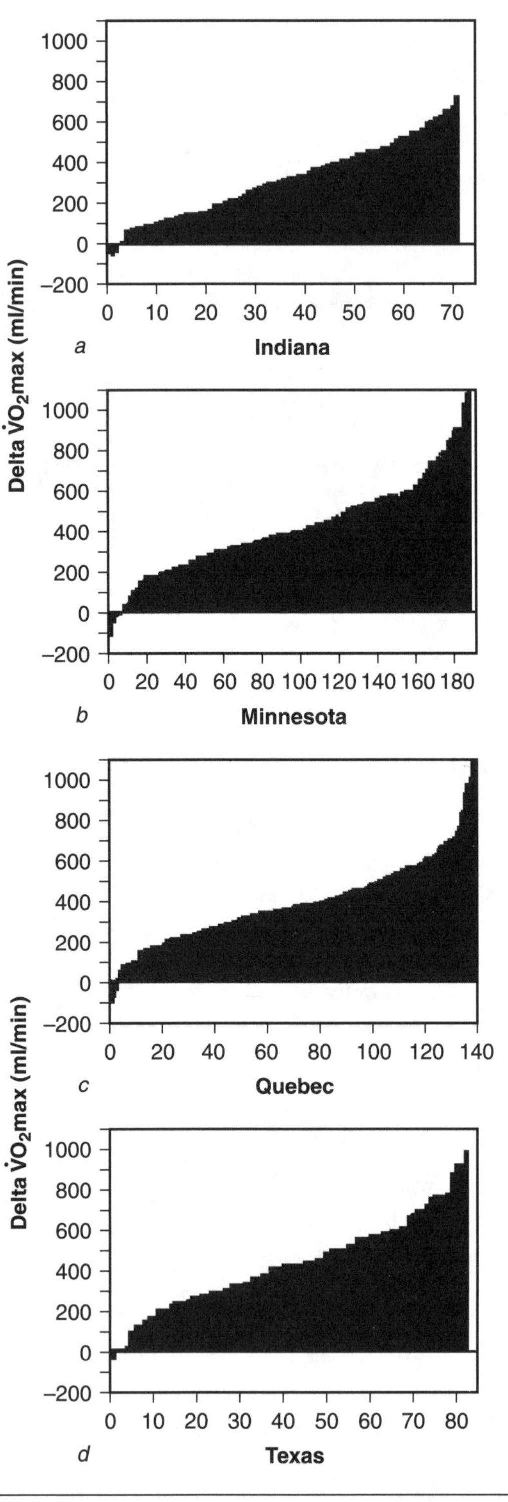

Figure 1.1 Individual differences in change (delta) in maximal oxygen uptake ($\dot{V}O_2max$) with exercise training for 481 individuals in the HERITAGE Family Study from four research centers: (*a*) Indiana, (*b*) Minnesota, (*c*) Quebec, and (*d*) Texas. The change in $\dot{V}O_2max$ is shown for each individual as a vertical bar, ordered from lowest to highest change. Note the considerable variability among individuals in each of the four research centers.

Reprinted from C. Bouchard et al., 1999, "Familial aggregation of $\dot{V}O_2max$ response to exercise training: Results from the HERITAGE Family Study," *J. Appl. Physiol.* 87(3): 1003-1008. Used with permission.

is definitely contradictory to the expected "normal" adaptation of $\dot{V}O_2max$ to such training. Moreover, the graphs also show that some participants at each of the four centers showed remarkably large improvements in $\dot{V}O_2max$ (i.e., much larger than expected from normative data), despite performing the *same* training intervention as those men and women who showed only minimal improvements. So, instead of all participants showing the typical 400 ml/min improvement, the actual range of observed values was from 0 ml/min to >1000 ml/min.

A logical question to pose in response to figure 1.1 would be, Is this extreme variability among individuals a typical phenomenon for other traits, or is it specific to $\dot{V}O_2max$? And the answer is that such extreme variability in the responses and adaptations of a large number of physical traits to exercise training is common. In an excellent review of this topic, Bouchard and Rankinen (2001) describe a number of physical traits that show quite variable responses to exercise training, including blood pressure, heart rate, and high-density lipoprotein (HDL) cholesterol. Figure 1.2 shows data from the HERITAGE Family Study for the response of HDL cholesterol to the 20-week exercise training program. If we were to examine historical, normative data showing the response of HDL (or "good") cholesterol to aerobic exercise training, we would expect individuals to show a slight improvement (increase) in HDL cholesterol levels as an adaptation to training. But, as seen in figure 1.2, while the average value for HDL cholesterol did increase slightly as expected, the individual adaptations of HDL cholesterol varied greatly, with many subjects actually having decreases in HDL levels. Even some individuals with low baseline HDL levels (shown in the left graph), who would be predicted to have the greatest likelihood of an increase in response to training, showed decreases or no change in HDL levels. So, to answer the question that began this paragraph: Variability among subjects in the response or adaptation to an exercise stimulus *is* normal, and the idea of an average or normative value is incomplete. Moreover, variability among individuals is common for nearly all traits, both at baseline and in response to an intervention or therapy; and examples from a wide variety of other areas could be included in this section, from medical or dietary interventions, with the same conclusion.

Another way to think about this issue is to consider the "normal curve," as shown in figure 1.3. For any physical trait that can be observed or measured (such a trait is also known as a **phenotype**), a broad range of values will likely be seen in a population of individuals, with the majority of individuals falling within a narrow range of values surrounding the **mean** or average value for the entire group. The breadth of this normal curve, or the extent of the **variation** observed between the lowest and highest

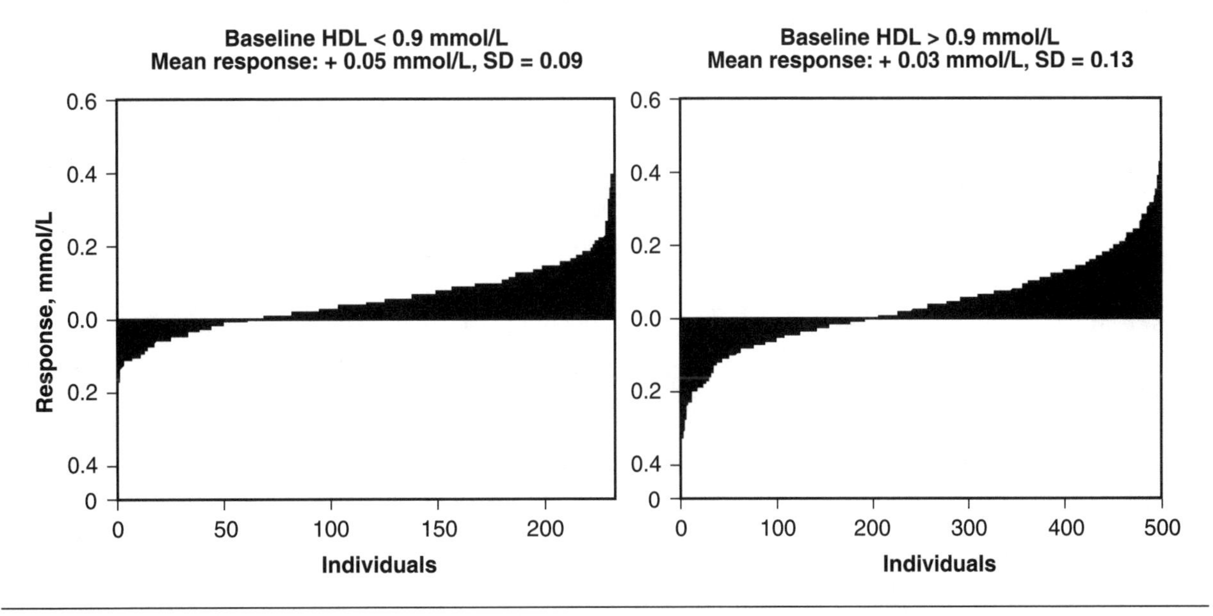

Figure 1.2 Individual differences in the response of HDL cholesterol to aerobic exercise training. Notice that both groups, regardless of baseline levels, included individuals who reduced HDL cholesterol in response to exercise training, even though the average response of each group was an increase in HDL level.

Reprinted, by permission, from C. Bouchard and T. Rankinen, 2001, "Individual differences in response to regular physical activity," *Med. Sci. Sports Exerc.* 33(6 Suppl): S446-451.

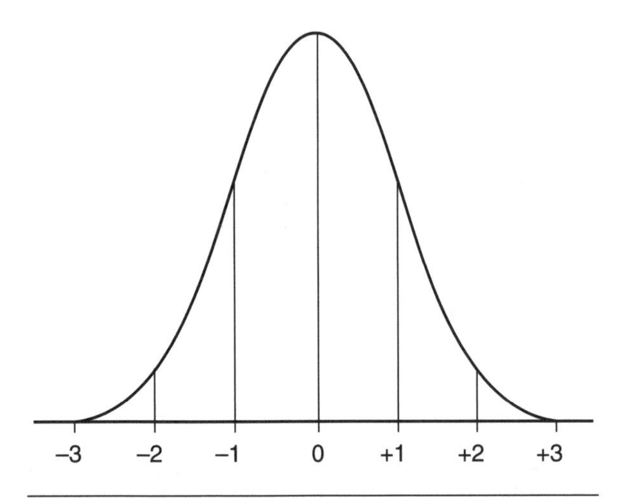

Figure 1.3 The standard normal curve, showing the mean or average value in the middle, with standard deviation values indicated (–1, +1, –2, etc.). The extremes of each end of the curve represent those individuals with the lowest and highest values for a particular trait, with all others falling in between those extremes.

values, is known as the *standard deviation,* which represents how much distance or variation lies between the extremes for a particular phenotype. The larger the standard deviation, or *variance,* surrounding a mean value, the greater the variability observed among different individuals for that phenotype measurement. As we've discussed from the beginning of this chapter, normative or average values that typify a response or adaptation to some stimulus are deceiving: The average value does not represent the value that all individuals have but rather the mean value, after averaging, of all the individual values spread across the extremes of the phenotype measurement (also known as **interindividual variability**). In research papers, this is represented by the standard deviation value or the variance underlying the mean value.

KEY POINT

The standard deviation value represents the extent of variability underlying an average value; larger standard deviations mean a greater range of individual values making up the overall average.

WHY DO WE SEE VARIATION IN A PHYSICAL TRAIT?

So a next logical question would be, What does this interindividual variation have to do with genetics? The answer to that question comes from addressing the *sources* of the variability that we see in individual values for a physical trait or phenotype of interest. In other words, what explains the great variability that we see in figures 1.1 and 1.2, and that is known to occur for so many other phenotypes? The sources of phenotype variability can be thought of as coming from three major factors: **experimental error, environmental factors,** and **genetic factors.** If we use $\dot{V}O_2$max as an example and refer back to figure 1.1, we are trying to understand what explains the great variability in how much $\dot{V}O_2$max improves or doesn't improve in response to aerobic exercise training. Let's examine the three sources of variability separately as we consider a research study that includes the measurement of maximal oxygen consumption.

Experimental Error

When beginning our research study, we will be interested in measuring our trait of interest as carefully as possible in order to test some hypotheses. For the phenotype of $\dot{V}O_2$max, recruited subjects will undergo a maximal oxygen consumption test during the research study. In order to improve the ability to test a particular hypothesis, we will want to carefully control this measurement to improve accuracy. In other words, we need to be aware of the accuracy and precision (i.e., validity and reliability) of the phenotype measurements. In the case of $\dot{V}O_2$max, an excellent study design can be wasted if the gas analyzers used for measuring oxygen consumption are poorly calibrated, or if the tubes needed to direct the flow of expired gases are faulty, and so on. Such problems result in variability in phenotype measurements: Even if the same person was measured many times, improper calibration or faulty equipment could lead to quite different values for each test despite similar performance by the person during each of those tests. This source of **error variability** is straightforward and can generally be controlled by adherence to established calibration and standard operating procedures. This goes for the measurement of any phenotype, be it cognitive function, blood pressure, enzyme activity, or whatever. Moreover, improvements in measurement devices are incorporated into research activities as technology improves and techniques are refined, resulting in continual reductions in this source of variability over time. For our purposes in this textbook, we will generally ignore measurement error as an important source of variability, because experimental error is typically small for most phenotype measurements.

Environmental Factors

Despite carefully calibrating our machines and using them correctly, we observe significant variability among our research subjects for the $\dot{V}O_2$max test. What differences among these individuals might explain some of this variability? Environmental factors are variables, such as exercise training history, body size, medical conditions or medications, diet, and room temperature, that are known to influence the physical trait or phenotype we're measuring. For example, we know that, on average, larger individuals will be stronger than smaller individuals, and exercise-trained individuals will have higher $\dot{V}O_2$max values than will sedentary individuals. A listing of key environmental factors will be specific to any particular phenotype, but the list represents those items that can be accounted for either in the research design or in the statistical analysis to reduce the variability seen in phenotype values across individuals. Researchers could control for these variables by studying, for example, $\dot{V}O_2$max in fasted older, postmenopausal, sedentary women in a climate-controlled setting at the same time each day. Such a research design reflects knowledge of the many environmental factors that could influence $\dot{V}O_2$max values, thus limiting the variability that researchers would expect in their measurement. We will spend additional time with this issue in chapter 9.

Genetic Factors

Even if we do an excellent job of measuring our phenotype accurately and accounting for (and controlling) known environmental factors, we will still see different $\dot{V}O_2$max values for the subjects of our study. In other words, measurement error and environmental factors may *not* completely explain all of the variability in the phenotype of interest (and they rarely will). The additional variability observed among individuals can then be accounted for by genetic factors specific to each particular person. Once the researchers have accounted for all known sources of experimental error and environmental factors, a key source of variation remaining has to do with each individual person's unique genetic makeup. Is there something unique to a given person's physiology that results in that person's unique phenotype response? In many cases, the answer to this question is yes. The focus of this textbook is on understanding what we mean by these "genetic factors" and discussing how genetics can be applied as a research tool in the areas of health, physical activity, and sport. We will begin our discussion by addressing the basics of genetics in the remaining chapters of part I.

KEY POINT

Individual variability in a phenotype (e.g., $\dot{V}O_2$ max) can be explained by experimental error, environmental factors, and genetic factors.

SUMMARY

While the use of average values is helpful for explaining typical phenotype values and their responses or adaptations to some stimulus, average values are often built on a foundation of extensive individual differences, as can be reflected by the normal curve. Such differences among individuals can be due to problems in measurement (experimental error), environmental factors not accounted for in study design, and individual genetic factors. Thus, in this chapter, we have learned the basic rationale for the study of genetics in various health- and fitness-related traits. Despite careful controls of environmental factors and measurement error, measurement values for many traits still differ considerably among different individuals, and genetic factors are likely to explain an individual's unique phenotype value. The remaining chapters are dedicated to understanding what is meant by genetic factors.

KEY TERMS

average
environmental factors
error variability
experimental error
genetic factors
interindividual variability
mean
phenotype
trait
variation

REVIEW QUESTIONS

1. What does an average value represent? What does the standard deviation value tell us about that average value?
2. What are the possible sources of the phenotype variability that is typically seen among individuals?
3. Pick a trait or phenotype of interest to you and identify specific environmental factors that would need to be accounted for or controlled in a research study of that phenotype.

CHAPTER

2

DNA, RNA, AND PROTEIN

Any journey begins with a preparation for basic needs, and our journey toward understanding genetics in health, physical activity, and sport is no different. In this chapter, we will develop our foundation of information by learning about the basics of our genetic material, DNA, and its place in the cell, as well as the roles of RNA and protein in the biology of the cell. As discussed in the Preface, this book is designed as an introductory text and as such should not substitute for a dedicated cell biology or biochemistry text for detailed information about the biology of DNA. That said, the information presented will be sufficient for all issues within this text and should provide readers with a strong foundation for moving forward with more advanced material.

BASIC CELL BIOLOGY

The cell is the basic functional unit of the body, with many different cell types specialized to perform specific functions for the various body tissues. Because of the various cell types and unique specializations, describing a "typical" cell is challenging. For example, muscle cells, neurons, and red blood cells are all cells, but their unique functional roles in the body result in remarkable structural differences. Nonetheless, a "generic" cell can be envisioned (figure 2.1). The cell has an outer cell membrane enclosing a *cytoplasm* that contains several organelles, including the endoplasmic reticulum, lysosomes, and the energy-producing mitochondria. The mitochondria (of which there

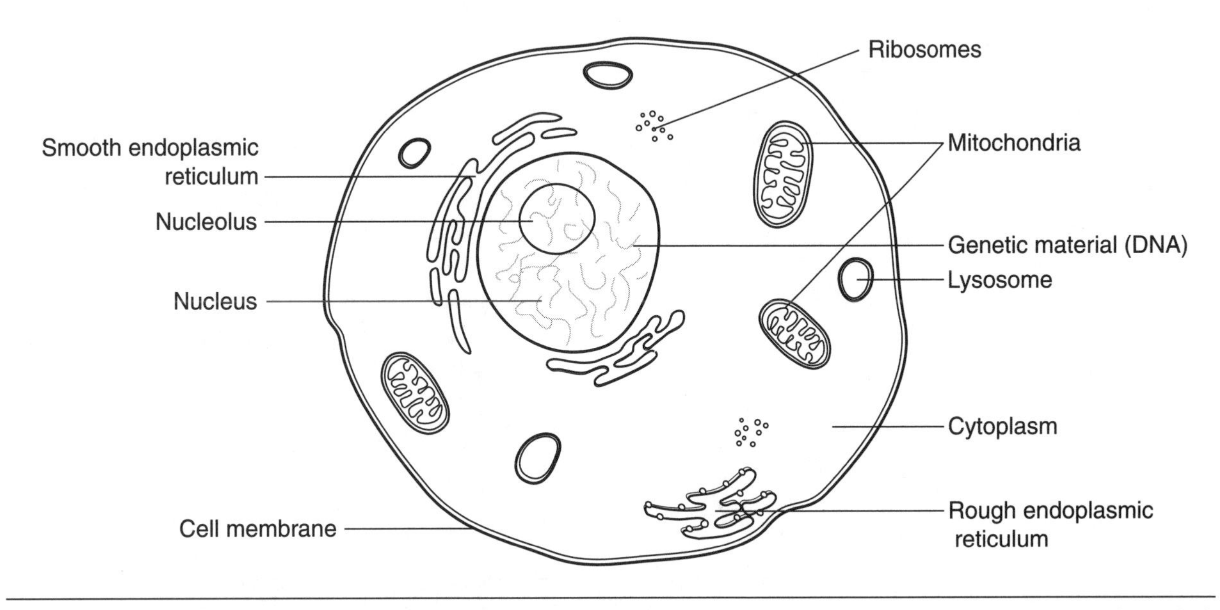

Figure 2.1 The "generic" cell, showing major components of all cells. The genetic material, DNA, is contained within the nucleus, which has its own nuclear membrane. Mitochondria have their own genetic material.

are hundreds to thousands within a cell) contain the enzymes critical for aerobic metabolism and are responsible for the majority of adenosine triphosphate (ATP) production in the cell. Within the cytoplasm, central to the cell, is a separate compartment, the *nucleus*, enclosed within its own nuclear membrane. The nucleus of the cell contains the **deoxyribonucleic acid,** or **DNA,** the instructional information of the cell. If the cell is thought of as a machine, the complete DNA sequence or **genome** contained in the cell is the instruction book for that machine. Remarkably, mitochondria have their own complement of DNA sequence (which is part of the *human genome*), which we will discuss in a moment.

DNA

The focus of this textbook is on genetics, which means we must focus on the instructional material of the cell, the DNA or deoxyribonucleic acid. DNA is a long polymer (i.e., chain of many molecules) composed of four nucleotide bases: **adenine, thymine, guanine,** and **cytosine,** commonly abbreviated as simply A, T, G, and C. These are the "letters" of the DNA sequence. DNA is a long string of these nucleotide bases or letters, the combination or order of which makes up the instructional information so important to the cell.

As shown in figure 2.2, DNA is more complex than simply a single string of nucleotide bases. In fact, DNA is a double, connected strand of **nucleotides**. One nucleotide base on one strand binds to a complementary base on the second strand, providing **complementary base pairs.** The nucleotide bases pair with each other in a specific, unchanging pattern: *A binds to T,* and G *binds to* C. Thus, information from one strand of DNA automatically provides information about the sequence of the second DNA strand. This complementary structure provides the foundation for making copies of DNA, both in the cell and in the laboratory.

The two strands of DNA, known as the "sense" and "anti-sense" strands or the "coding" and "template" strands, form a twisting ladder structure known as the double helix. The complementary binding of A to T and G to C means that each rung of the DNA ladder represents a chemical bond between two complementary bases (e.g., A binding to T). This double-stranded arrangement provides a means to make exact copies of DNA, thus preserving genetic information during the process of **DNA replication** and cell division. Although the sequence of DNA appears random, the reality is that these long strings of apparently random letter sequences actually correspond to specific instructions for the building of various cell components, primarily proteins. These regions of DNA sequence that provide specific instructions for making proteins and other cell components are known as **genes** or **gene regions.** The **human genome,** or the entire complement of human DNA sequence, contains 20,000 to 25,000 genes.

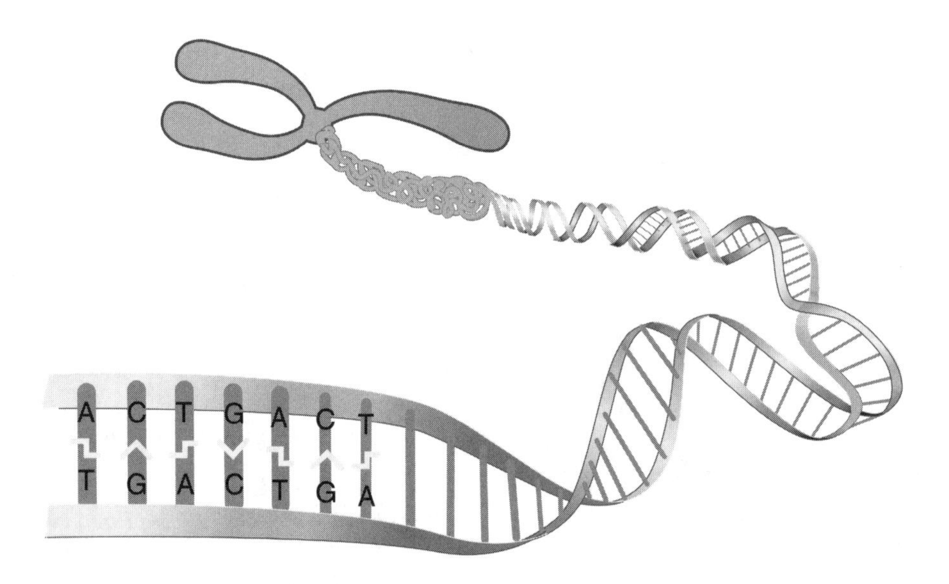

Figure 2.2 The basic structure of DNA. The two complementary strands have a double helix structure. DNA is contained in chromosomes within the cell nucleus.

KEY POINT

The human genome consists of 3.1 billion nucleotide letters (A, T, G, and C), which contain 20,000 to 25,000 gene sequences.

OUR GENETIC MATERIAL AS CHROMOSOMES

Although our DNA or genome can be thought of as a long string of letters (i.e., nucleotide bases), the DNA of the genome is in fact separated into unique components known as **chromosomes.** The genome of each person is composed of 23 paired chromosomes (one set of 23 from the mother, and one set of 23 from the father) for a total of 46 chromosomes, each of which contains a unique string of DNA sequence. All 46 chromosomes are contained within the nucleus of each cell (an exception being red blood cells, which have no nucleus or DNA). Twenty-two of these chromosomes, numbered 1 through 22, are referred to as **autosomes.** Each cell contains two copies of each autosome. The **sex chromosomes,** X and Y, carry the DNA sequence and genes that relate primarily to sexual differentiation, with the Y chromosome carried only by males and carrying genes important for the development of the male sex organs. Thus, of the 46 total chromosomes, the cells in women carry two copies of each autosome plus two copies of the X chromosome. In contrast, the cells in men carry two copies of each autosome plus one X chromosome (from the mother) and one Y chromosome (from the father).

The exception to this discussion of chromosomes is mitochondrial DNA, which is separate from the DNA contained within the cell nucleus. Mitochondrial DNA (sometimes abbreviated mtDNA) is a small, 16,568-nucleotide genome contained in all mitochondria. Rather than a linear molecule of double-stranded DNA, mitochondrial DNA is a circular double helix containing 37 genes; most of these genes are specific to the manufacture of new mitochondria, as would occur, for example, in skeletal muscle as an adaptation to aerobic exercise training. With rare exceptions, mitochondrial DNA is inherited solely from the mother; thus the mother and her offspring will share the same mitochondrial DNA sequence. For our purposes, we can think of the mitochondrial genome as just another part of the human genome: Its genes are just as important to cell biology as the genes located in the cell nucleus, and the processes and rules that govern nuclear DNA are very similar to those for mitochondrial DNA.

Each of the chromosomes in the body contains a distinct DNA sequence with a specific set of genes that is arranged in an unchanging order along its length, and all humans share this chromosome geography. In other words, all humans, regardless of age or race, share the same chromosome structure and the same complement of genes. We'll see in subsequent chapters why it is that each individual is uniquely different (excepting perhaps identical twins). The sexes differ by the presence or absence of the Y chromosome or, alternatively, the total number of X chromosomes (two for women, one for men).

KEY POINT

Each cell nucleus contains 46 chromosomes: two copies of each of the 22 autosomes and 2 sex chromosomes (X and Y for men, two X chromosomes for women).

In the early days of genetics research, the exact sequence structure of DNA was very difficult to determine, and researchers instead focused on studying the general structure of chromosomes. Chromosomes are actually more than just the DNA sequence. In addition to DNA, chromosomes are composed of various proteins used to stabilize and package the DNA in the cell nucleus. Thus, chromosomes act as a structural backbone for our DNA material, allowing for expansion of the DNA sequence when its information is needed by the cell, and alternatively allowing highly compact packaging of the DNA during the process of cell division.

In its tightly packed arrangement, each of the chromosomes takes on a unique appearance, allowing the identification of each of the autosomes and sex chromosomes from each other. In fact, each chromosome has a unique and unchanging structural identity, often referred to as a *banding pattern* (different parts of the chromosome appear as light and dark bands under certain laboratory conditions). These unique chromosome banding patterns have long been used as a means to locate different regions of DNA, thus providing "signposts" or a geography for locating specific genes.

With modern genetics laboratory techniques, the use of the chromosome banding pattern as a means of locating genes is unnecessary; however, the bands are still used for describing the location of specific genes. Each chromosome has a unique geography of patterns, and a specific nomenclature is used to describe each band along its length. Thus, each band has its own label, and genes are often described as being located at a specific band within a specific chromosome. This location is referred to as a **locus** (the plural of which is **loci**). Figure 2.3 provides an example. The typical format for a band label uses three elements, as shown in the label 12p13. The first number, 12, is the chromosome number (1-22 or X or Y). The letter immediately following the chromosome number describes the major region of the chromosome, known as the short or the long arm of the chromosome. The letter *p* is used to denote the short arm (from the French, "petit," for "small"), and the letter *q* denotes the long arm (*q* for "queue"). Finally, the numbers following the *p* or *q* correspond to the specific band on the chromosome. Thus, the band pattern label is much like a home address: From it you can find the specific chromosome, the major region of

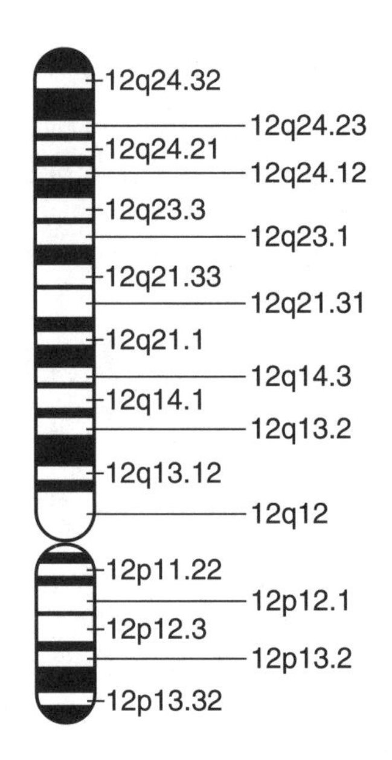

Figure 2.3 An example of a chromosome banding pattern and nomenclature, shown for human chromosome 12.

From the Genetics Home Reference, National Library of Medicine (US). http://ghr.nlm.nih.gov

SPECIAL FOCUS
The Human Genome Project

In the mid-1980s, an idea was proposed—to determine all of the letters of the human genome, to completely sequence the nucleotide bases that make up human DNA. The idea was controversial at the time, given the enormous technical challenges and cost required to identify the more than 3 billion letters in the genome; but the anticipated medical benefits of understanding the entire DNA sequence made for a stronger counterargument. Over several years, the Department of Energy (DOE) and National Institutes of Health (NIH) began funding pilot projects directed at developing the necessary technologies and other resources needed to begin such an ambitious undertaking, and by the late 1980s, the U.S. Congress came on board to support the effort. The U.S. Human Genome Project (HGP) began formally in 1990 (see its logo in figure 2.4). While the project was initiated within the United States, the work became international and was eventually completed by a close collaboration of 20 research centers in six countries around the world.

Through the mid-1990s, maps of the human chromosomes were completed and small regions of the genome were sequenced. A major difficulty

Figure 2.4 The Human Genome Project logo.

From the U.S. Department of Energy Human Genome Program. http://www.ornl.gov/hgmis

with sequencing the entire genome was that a limitation of the sequencing technology prevented sequencing more than 500 to 700 nucleo-

(continued)

SPECIAL FOCUS

(continued)

tide bases at a time, which meant that sequencing the entire genome required the splicing together of millions of small fragments of DNA sequence. Thus, intensive bioinformatics (i.e., computer and statistical) resources were also developed to aid in the aligning and matching of adjoining sequence strands. By 1997, the NIH established the National Human Genome Research Institute (NHGRI) in recognition of the growth of the sequencing efforts and to enhance collaboration with other NIH institutes. In 1998, the DOE and NIH submitted a new five-year plan to Congress; the revised plan was to complete the entire sequence of the human genome by 2003, two years earlier than originally envisioned, with a draft sequence (90% complete, but still quite useful for researchers) anticipated in 2001. At this same time, Celera Genomics, a private company, announced plans to sequence the entire genome independent of the HGP efforts within three years. Advances in sequencing technology, namely capillary-based sequencers, significantly improved the speed of sequencing while reducing costs, resulting in strong progress for both the public and private efforts. In June 2000, President Bill Clinton, joined by Celera Genomics President Craig Venter and NHGRI head Francis Collins, announced the completion of the draft sequences of the human genome, both versions of which were published in early 2001 (International Human Genome Sequencing Consortium, 2001; Venter et al., 2001).

Following the publication of the draft sequences, the HGP continued its work, refining and completing the entire genome sequence. In 2003, very near the 50th anniversary of the identification of the double helix structure of DNA by James Watson and Francis Crick, the HGP announced the completion of the final human genome sequence, with 99% coverage of the 3.1 billion base pairs at 99.99% accuracy. This final version came two years ahead of the original schedule at a cost lower than originally anticipated ($2.7 billion in 1991 dollars).

The efforts of the NHGRI did not end with this final genome sequence. As will be discussed in other parts of this textbook, simply having the sequence of the genome does not provide insight into its function, nor does it provide answers to the complexity of human biology. The NHGRI remains committed to furthering research aimed at understanding the complexity and underlying function of the genome, including genetic sequence variation with likely implications for individualized or genomic medicine. Additional information can be found in Collins et al. (2003) and online at www.doegenomes.org.

the chromosome, and the specific band in that region in which your gene of interest is located.

RNA

The DNA sequence is simply information for the cell, and the basis for all cellular function falls upon two other molecular machines in the cell, namely RNA and protein. **Ribonucleic acid (RNA)** has the primary job of carrying the DNA information from the cell nucleus to the cell cytoplasm, where proteins are manufactured. RNA is different from DNA in that it is composed of a single strand of nucleotides (rather than two strands for DNA) and contains the nucleotides adenine (A), guanine (G), cytosine (C), and **uracil** (U). Uracil replaces the DNA nucleotide thymine (T) in the RNA sequence and similarly binds only to its complementary nucleotide, adenine (e.g., U binds to A). This *complementary binding* of nucleotides is necessary for the DNA information to be accurately coded into RNA.

As discussed in detail in the next chapter, **transcription** is the process of reading DNA and coding it into a complementary RNA strand. Because of the specific binding constraints of the DNA and RNA nucleotides (G with C, and A with T/U), the DNA sequence can be easily used to manufacture a complementary RNA sequence without loss of information. The information for each gene is transcribed into an RNA sequence from the DNA sequence. Once built, the final RNA molecule is free to move to the cell cytoplasm where it can be used for the manufacture of a protein molecule. Often, multiple copies of the RNA molecule will be built, depending on the needs of the cell.

While there are multiple types of RNA that can be produced by the cell from the DNA sequence, the

RNA of focus for our purposes is that which is used for making proteins. This RNA is given the special name of **messenger RNA (mRNA)** because of its movement from the cell nucleus to the cell cytoplasm. In other words, the RNA acts as a messenger, carrying the complementary DNA sequence information from the cell nucleus to the site of protein manufacture in the cell cytoplasm.

AMINO ACIDS AND PROTEINS

Proteins perform the bulk of the work of any cell, be it structural or functional (e.g., enzymes). Proteins are molecules composed of a specific linear sequence of building blocks known as amino acids. Similar to the nucleotides in DNA and RNA, proteins are simply strands of linked amino acids. These amino acid chains ultimately take on complex three-dimensional structures specific to each protein. Twenty amino acids are used in the manufacture of proteins within cells, and these amino acids are specifically coded by the DNA and RNA sequence (table 2.1). Amino acids are also known as *peptides;* thus proteins are sometimes called *polypeptides.*

A cell must continually respond to its particular environment in order to maintain homeostasis. Often, these responses require the manufacture of new proteins or additional copies of existing proteins in the cell (imagine the need for more contractile proteins in hypertrophying skeletal muscle). The cell relies on the information coded within the DNA to build these proteins, using the mRNA molecule as an intermediate messenger system. The DNA information for the specific protein is first coded into mRNA within the cell nucleus, after which the mRNA moves to the cell cytoplasm where it is acted upon by the translation machinery of the cell to make the resulting protein sequence. **Translation,** performed by cellular machines known as ribosomes, is the process of reading mRNA sequence and building the complementary amino acid strand. This is shown generally in figure 2.5.

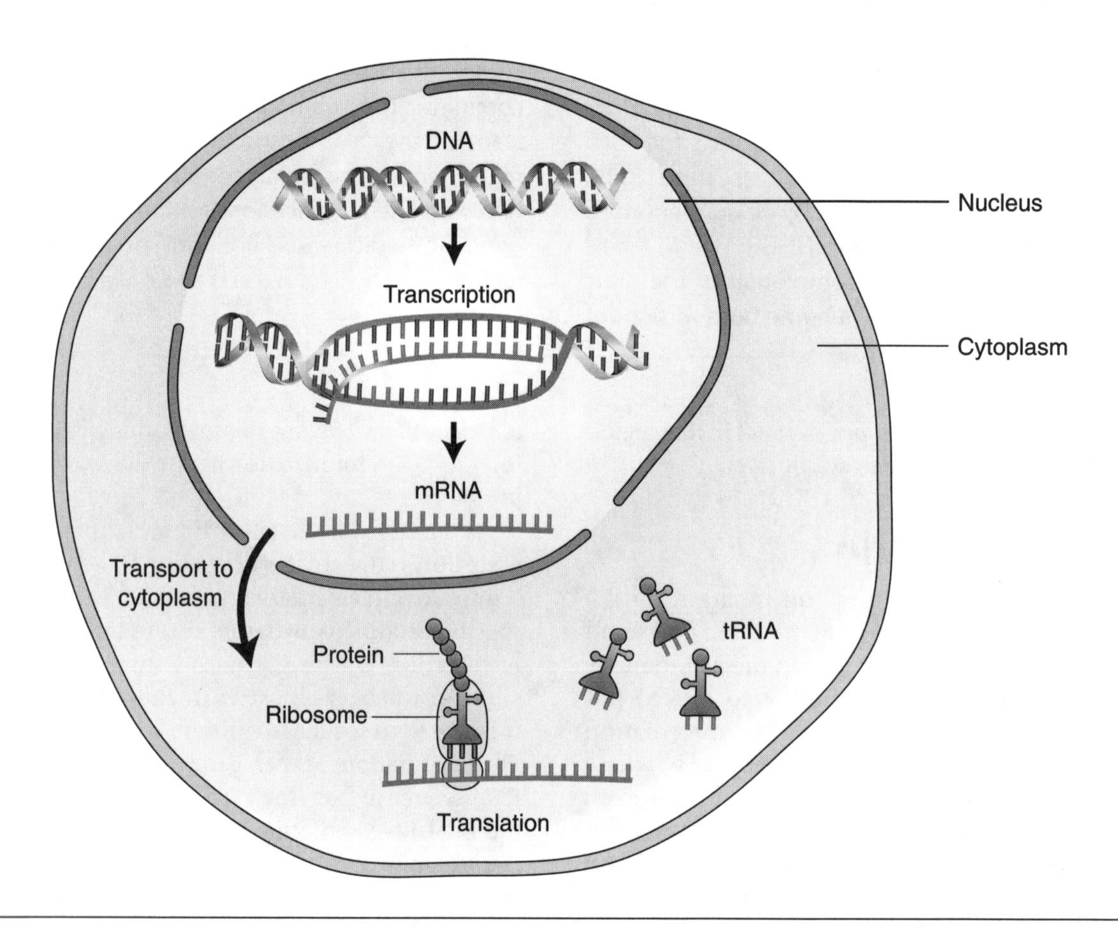

Figure 2.5 The basic processes of reading DNA sequence information and using it to build a protein molecule. Notice that the DNA sequence is read in the cell's nucleus, where a complementary RNA strand is built. That mRNA strand then moves to the cell cytoplasm, where it is used to manufacture the amino acid sequence of the protein.

Table 2.1 The 20 Amino Acids and Their Three-Letter and Single-Letter Abbreviations

Amino acid	Three letters	One letter
Alanine	ALA	A
Arginine	ARG	R
Asparagine	ASN	N
Aspartic acid	ASP	D
Cysteine	CYS	C
Glutamic acid	GLU	E
Glutamine	GLN	Q
Glycine	GLY	G
Histidine	HIS	H
Isoleucine	ILE	I
Leucine	LEU	L
Lysine	LYS	K
Methionine	MET	M
Phenylalanine	PHE	F
Proline	PRO	P
Serine	SER	S
Threonine	THR	T
Tryptophan	TRP	W
Tyrosine	TYR	Y
Valine	VAL	V

STRUCTURE OF GENES

Although the human genome totals 3.1 billion DNA nucleotides (A, T, G, and C), not all of those nucleotides contain information for making proteins. The information packets within DNA used for making or "coding for" proteins are known as genes. In other words, a gene is a stretch of DNA sequence that provides the information necessary to build specific proteins. Somewhere between 20,000 and 25,000 genes are estimated to exist within the human genome; the complete list is still being determined. While some of these DNA regions code for material other than protein (e.g., structural RNA molecules), for our purposes a gene is a specific stretch of DNA sequence that codes for one or more proteins (as we'll see, different parts of a single gene can be used to build multiple proteins, thus allowing more than 20,000-25,000 proteins in the body). The thousands of genes are found across all of the 22 autosomes, the X and Y chromosomes, and the mitochondrial genome, in locations specific to each gene.

As shown in figure 2.6, the generic gene structure consists of three general regions required for gene function: a beginning and an end, with information critical to manufacturing the protein found in different parts of the middle region. The beginning or starting region of a gene is known as the **promoter** or promoter region. This is also known as the *upstream* region of the gene. In general, the promoter region contains the signals necessary for turning a gene on or off—signals critical for starting or stopping the process of transcription of DNA into RNA. In reality, the gene's promoter region is far more complex, containing many signal sequences that influence the machinery that reads DNA sequence and builds the complementary RNA strand. These signals within the promoter are small stretches of DNA sequence, several nucleotides in length, that are recognized by the proteins important to the transcription of that gene. These **transcription factor proteins** (or simply *transcription factors*) bind to their specific signal sequences in the gene's promoter and, depending on their specific function, work to

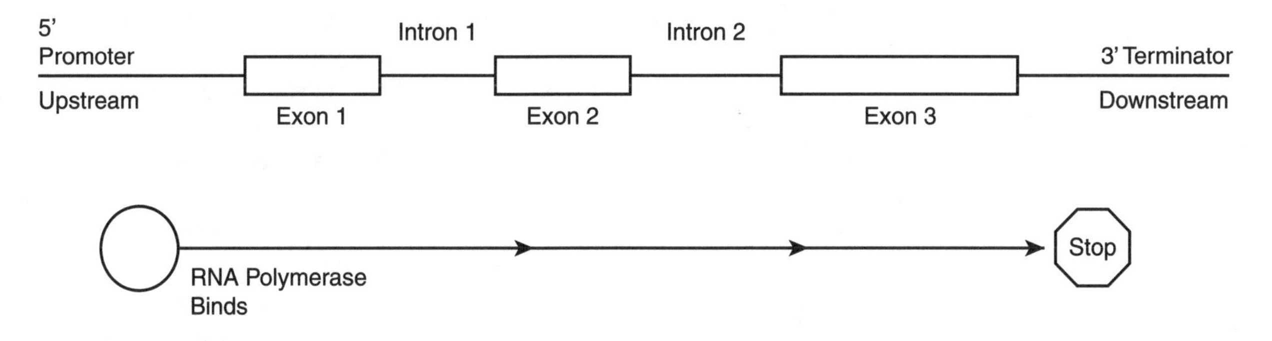

Figure 2.6 The generic structure of a gene, including the upstream promoter region, the coding exons separated by introns, and the downstream region at the end of the gene. The RNA polymerase enzyme binds to the promoter region, reading the DNA sequence until just past the final exon, resulting in the production of a complementary RNA strand.

promote or inhibit the transcription machinery, thus regulating the transcription of the gene. One way to think of this is that the promoter acts on the gene much as a dimmer switch acts on a lightbulb: The switch can be on or off, and if it is on, the brightness can be adjusted from low to high. In the case of the promoter region, the gene is considered turned on when RNA is being manufactured from the DNA (also known as transcription or gene expression), and the promoter region also regulates how much RNA is manufactured.

The promoter region is the target for the various proteins that regulate and transcribe the gene. That transcription process begins just before (or just upstream of) the region that contains the information important for coding the protein, the middle region or **coding region** of the gene. This coding region is made up of two major components: exons and introns. **Exons** are stretches of DNA sequence that contain information critical to the protein's amino acid sequence, while the **introns** are stretches of DNA sequence between exons that do not have information important to the protein sequence. In other words, the entire DNA sequence that codes for a protein is broken up by stretches of noncoding DNA sequence. In the process of transcription, the entire coding region of DNA (exons + introns) is copied into the mRNA; however, the intronic or noncoding regions are removed from the mRNA sequence before the protein is made (as will be discussed in chapter 3). The coding region of the gene can contain anywhere from one exon, with no introns, to many exons, each separated by an intronic sequence. The exon and intron sequences can be fewer than 100 bases in length or can be thousands of bases in length. The transcription process, the coding and manufacture of the RNA sequence, begins just before the first exon and continues until just past the final exon.

KEY POINT

Genes consist of three regions: the promoter or upstream region, the middle or coding region, and the terminator or downstream region. The information critical to the amino acid sequence is found in the coding region.

One quick note about introns: They are not wasted space within the gene. While the importance of introns is not fully known, these regions separating exons can contain signal sequences similar to those of the promoter region. The intronic regions can bind proteins that influence the regulation of the gene. Moreover, the introns contain information needed for their removal from the mRNA sequence before the protein is manufactured.

The end of the gene is similar to the beginning of the gene in that it also contains signaling regions important to regulating gene transcription. The end of the gene, also known as the **terminator region** or the *downstream* region, contains the information important to stopping the transcription machinery and thus defining the end of the mRNA molecule. Also similar to the promoter region, the end of the gene can contain various signaling sequences that attract proteins; these can positively or negatively influence the transcription of the gene.

Figure 2.7 provides a graphic version of the major information presented in this chapter. It gives an overview of the structure of DNA from the chromosome level to the exons and introns within a gene. Also shown is the process of transcription of a gene to mRNA and translation of that mRNA to protein.

The drawing on the left represents a condensed chromosome, with its banding pattern indicated. These bands are used as geographical or anatomical landmarks for the chromosome and can be used to indicate the general location of a specific gene. Each chromosome contains millions of nucleotide bases or DNA letters, with hundreds to thousands of different genes. This is followed by a small length of the chromosome, with four genes indicated as short lines next to the long strand of DNA. Each gene would produce an independent gene product, such as a unique protein. Third, a portion of one of the genes

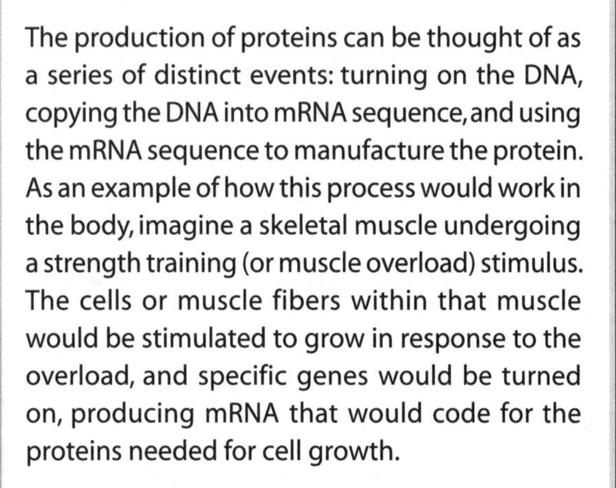
KEY POINT

The production of proteins can be thought of as a series of distinct events: turning on the DNA, copying the DNA into mRNA sequence, and using the mRNA sequence to manufacture the protein. As an example of how this process would work in the body, imagine a skeletal muscle undergoing a strength training (or muscle overload) stimulus. The cells or muscle fibers within that muscle would be stimulated to grow in response to the overload, and specific genes would be turned on, producing mRNA that would code for the proteins needed for cell growth.

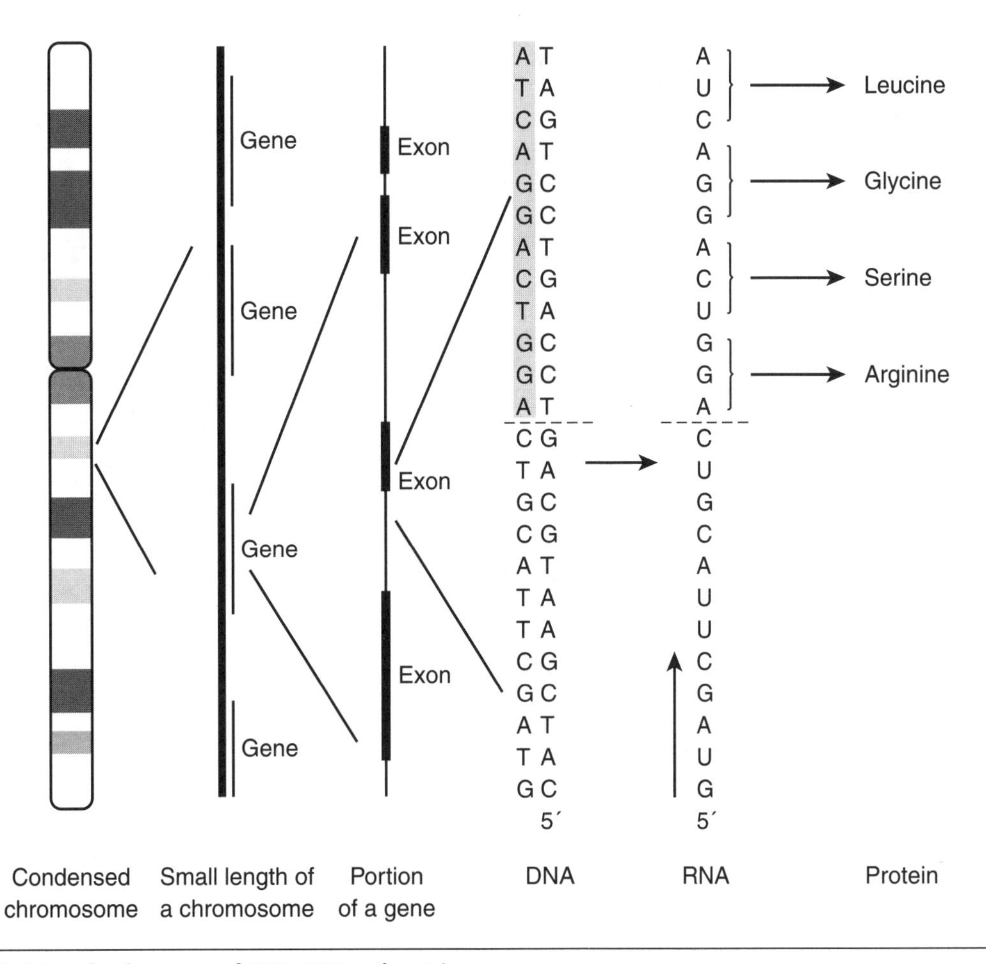

Figure 2.7 The hierarchical structure of DNA, RNA, and protein.

is shown: We see the DNA strand as a long thin line, with the gene exons shown as dark bands, separated by the thin introns. In the DNA section, we see a small fragment of the gene sequence, including part of an intron and part of an exon; the exon–intron boundary is indicated by the dashed line. This DNA sequence is then transcribed (i.e., *transcription*) to RNA, whereby each DNA base is used to build a complementary RNA strand (with T replaced by U in the RNA alphabet). Finally, the exonic regions of the DNA are used for the instructional material needed to translate RNA into amino acids and a complete protein (i.e., *translation*). Each three-letter sequence within the exons corresponds to a specific amino acid. The amino acids are linked together by the ribosome to build the protein's amino acid chain. Many of the processes shown in figure 2.7 will be discussed in chapter 3.

SUMMARY

The functioning cell is a marvel of complexity, with multiple structural and functional systems all working in concert to maintain cell (and thus tissue and organism) homeostasis. Proteins perform the work of the cell, but the manufacture of those proteins requires the information stored within DNA. The DNA's sequence of As, Ts, Cs, and Gs is stored within the cell nucleus in the form of chromosomes. Within each chromosome, specific regions of DNA known as genes code for proteins through the messenger RNA. These genes are regulated by a number of transcription factors that bind to regulatory signals in the promoter and other regions of the gene, thereby influencing the extent to which the genes are turned on. The resulting mRNA then travels to the cytoplasm, where it is used to manufacture the coded protein.

KEY TERMS

adenine
amino acid
autosome
chromosome
coding region
complementary base pairs
cytosine
deoxyribonucleic acid (DNA)
DNA replication
exon
gene
gene regions
genome
guanine
human genome
intron
loci
locus
messenger RNA (mRNA)
nucleotide
promoter
protein
ribonucleic acid (RNA)
sex chromosome
terminator region
thymine
transcription
transcription factor proteins
translation
uracil

REVIEW QUESTIONS

1. Where is DNA located in the cell and how is it organized? How does this organization differ between males and females?
2. What are the basic components of DNA and RNA? How are DNA and RNA different?
3. Describe the general process by which DNA is used to manufacture a protein.
4. What parts of a gene contain coding information for the amino acid sequence in a protein? What are the parts of the gene that contain regulatory sequences important for gene function?

CHAPTER

3

TRANSCRIPTION, TRANSLATION, AND THE GENETIC CODE

Now that the basic structures of DNA biology have been described, we begin our discussion of how these molecules interact, all of which will have relevance for how genetic factors influence phenotypes, or traits. This chapter presents a brief overview of the basic mechanisms by which DNA is read and ultimately used to manufacture an amino acid sequence.

MOVING FROM DNA TO RNA: TRANSCRIPTION

The "instructions" of DNA are carried out within the cell by proteins, which act as the primary structural and functional units of the body. Just as any instruction manual must be interpreted by the reader before a complex task can be accomplished, in order for the genetic information encoded within DNA to be converted into proteins, the DNA must be read and interpreted. This is a complex task, which begins in the cell nucleus where DNA is stored. The DNA is converted to the complementary structure mRNA by a process known as **transcription** or **gene expression.** Transcription is the process whereby the genetic material in DNA is read by cell machinery known as **RNA polymerase** and used to generate a messenger RNA molecule. As described in chapter 2, the promoter region of a gene contains signaling sequences that are identified by the RNA polymerase enzyme and other transcription factors. When a gene is targeted for transcription (meaning that the cell has been signaled somehow to produce more of a particular protein), the RNA polymerase binds to the promoter region and begins to move from the beginning (the upstream region) to the end of the gene (the downstream region). As each DNA nucleotide is read by the polymerase enzyme, the corresponding RNA nucleotide is assembled in series on the growing mRNA chain. This process continues until the polymerase encounters the terminator region, which is found past the final exon of the gene. After this transcription process is finished, the mRNA molecule moves out of the nucleus and into the cell's cytoplasm. It is this movement of the mRNA out of the nucleus that gives it the name "messenger RNA."

KEY POINT

Transcription, or gene expression, is the process by which the RNA polymerase enzyme reads the DNA sequence and manufactures a corresponding mRNA sequence.

As described in chapter 2, the mRNA molecule's sequence is an exact duplicate of the DNA sequence, with one exception. The nucleotide thymine (T) is not found in mRNA; rather uracil (U) is the base complementary for adenine (A) in mRNA. Thus, for any DNA coding sequence ATCG found in the genome, RNA polymerase would transcribe this gene region into the corresponding RNA sequence of AUGC.

POSTTRANSCRIPTIONAL MODIFICATIONS OF RNA

Once the mRNA molecule is completed, it is targeted for modification before the process of translation can begin. The goal of the mRNA molecule is to provide information for the generation of a protein, but recall from chapter 2 that not all of the DNA sequence is useful for protein manufacturing. Only the exons within a gene are used for generating an amino acid sequence; so the mRNA, which is a complete duplicate of the entire gene region, contains more information than is useful because it retains the intronic sequences of the gene. Thus, the mRNA requires some modification before it can be used as a template for protein generation. This process is known as **posttranscriptional modification,** as it is occurring after transcription of the RNA molecule.

As shown in figure 3.1, the mRNA molecule initially contains all of the information coded for in the exons and introns of a particular gene. But before the manufacture of a protein (i.e., before translation), the intron sequences are removed from the mRNA strand, resulting in just the exonic regions being retained in the mRNA sequence for making the final protein. The final mRNA sequence corresponds to just the coding, exonic regions of the DNA. This key event in posttranscriptional modification is a process known as **splicing,** whereby the intronic sequences are literally cut out of the RNA molecule and the two free ends of each neighboring exon are "spliced" together.

A key modification to mRNA structure that we should address is **alternative splicing.** In the human genome, there are 20,000 to 25,000 genes but more than 100,000 proteins. Where are these additional proteins coming from? Many genes produce more than one protein through a process known as alternative splicing. In this process, certain exonic sequences are removed from the mRNA molecule, similar to the process for removing intronic sequences. In other words, not all of the exons for any particular gene are required for a particular protein, and different combinations of exonic sequences can be used to produce unique proteins. This process is shown in figure 3.2.

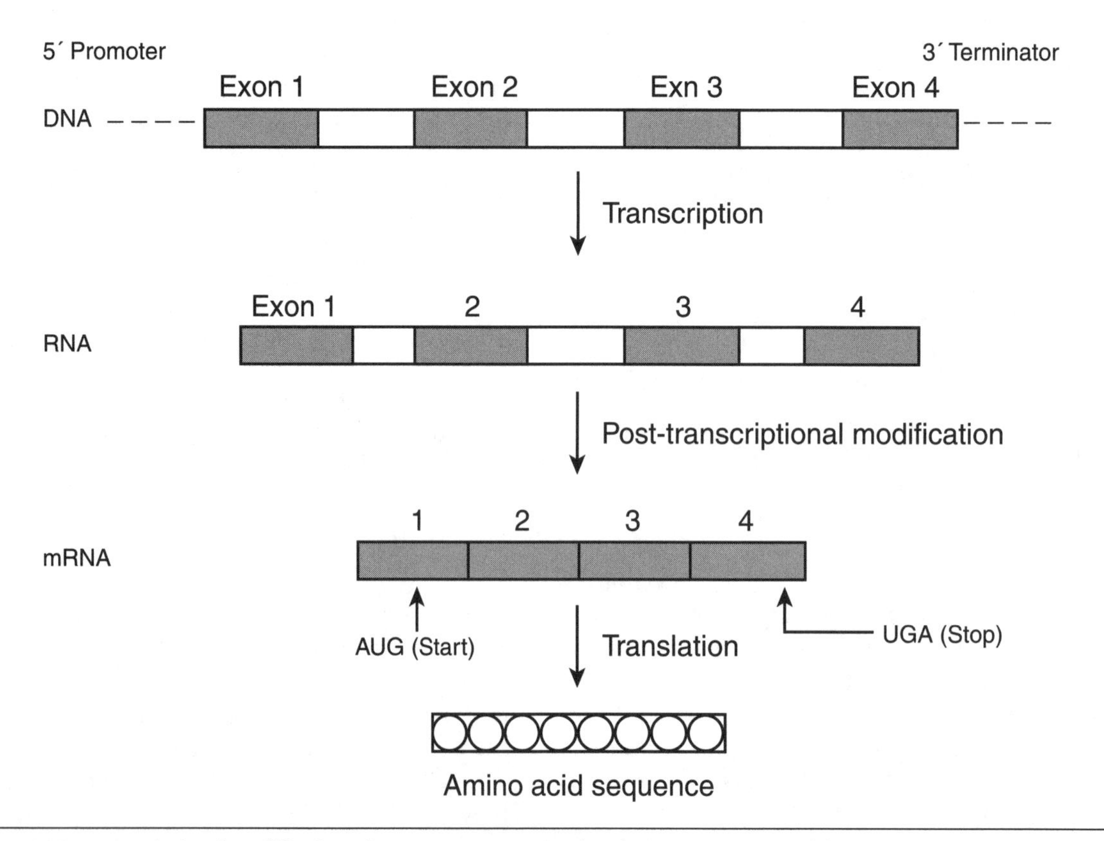

Figure 3.1 Posttranscriptional modification of RNA sequence. Notice that the intronic regions of the DNA sequence are removed from the RNA sequence prior to the translation into amino acid sequence.

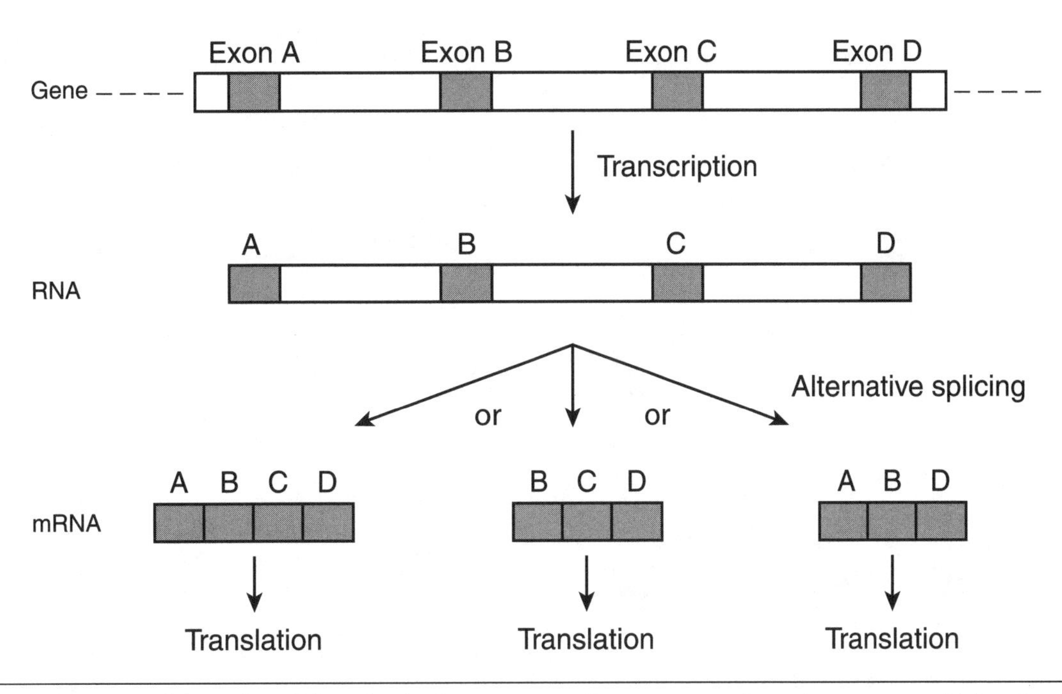

Figure 3.2 The process of alternative splicing, whereby the same gene coding sequence can be used to generate multiple, unique proteins through the shuffling or deletion of specific exons.

MOVING FROM RNA TO PROTEIN: TRANSLATION

Now that the mRNA molecule has been trimmed of its excess sequence information, it can provide the template for the encoded protein or amino acid sequence. This process is performed by the **ribosome,** a complex structure within the cell composed of several molecular parts. The ribosome recognizes the 5' end (i.e., the sequence beginning) of the modified mRNA molecule and begins a process known as translation. **Translation** is the act of reading the three-letter sequence combinations within the mRNA sequence and using that information to build the corresponding amino acid sequence according to the genetic code (described at the end of the chapter). In other words, the ribosome moves from the beginning of the mRNA molecule to the end, reading each adjacent three-nucleotide sequence to build the corresponding protein. Each of these three-letter sequences is known as a **codon.**

The ribosome begins its work at the **initiation codon** or **start codon,** or a series of three nucleotide bases that are universally recognized by the cell as the first codon for making the protein. This initiation or "start" codon is AUG in the mRNA sequence and corresponds to the amino acid methionine (MET) according to the genetic code. What this means is that all proteins in the body begin their lives as a methionine amino acid, upon which other amino acids are added in sequence by the ribosome. As shown in figure 3.3, the ribosome will continue to read all of the neighboring codons and build the corresponding amino acid chain until it reaches a specific stop signal. Within the genetic code, there are three stop signals, known as **termination codons** or **stop codons,** recognized by all ribosomes: UGA, UAA, and UAG. Once the ribosome recognizes a stop codon, it disengages from the mRNA sequence and releases the finished amino acid sequence. The region in between the start and stop codons is known as the *open reading frame,* as the ribosome will freely read the mRNA sequence along this length until it reaches a stop codon. The start and stop codons are typically located near the beginning of exon 1 and the end of the last exon, respectively, though in some cases considerable lengths of exonic sequence fall outside of this open reading frame.

KEY POINT

Translation is the process by which the ribosome reads the codons within the mRNA sequence and builds the corresponding amino acid sequence.

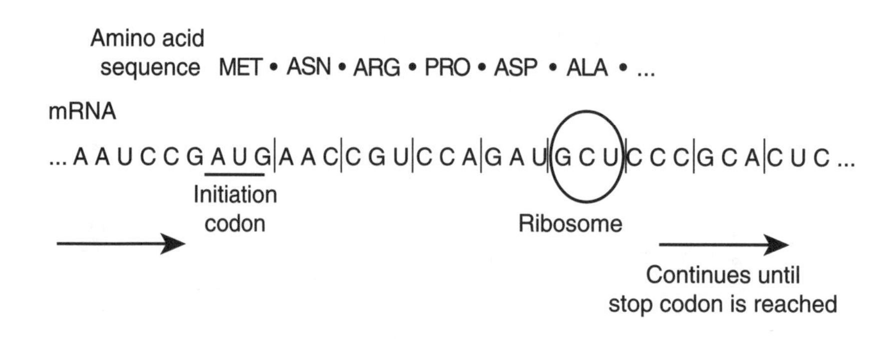

Figure 3.3 The building of an amino acid sequence by reading of codons. The ribosome, beginning at the initiation codon, reads each triplet codon in series, building the corresponding amino acid chain, and stopping when it reaches a termination codon.

POSTTRANSLATIONAL MODIFICATIONS OF PROTEINS

The process of translation provides an amino acid sequence that corresponds exactly to the mRNA molecule that was modified after first leaving the nucleus, which means that it corresponds to all or part of the original coding regions of the DNA sequence for that gene. The completed amino acid sequence will then typically undergo a series of **posttranslational modifications** in order for that protein to function perfectly. Several such modifications are possible, though perhaps the most important is the folding of the amino acid sequence into its final three-dimensional shape, which is required for protein function. This is often accomplished by the interaction of the amino acid sequence with other proteins, known as chaperones. Moreover, many proteins function in direct combination with other proteins, so these partners will have to be bound to the original protein sequence.

Once a protein is manufactured and folded, it may be required to move back into the nucleus or even be released from the cell altogether. Such movement of proteins across cell membranes requires signals, which are in the form of specific amino acid sequences that are recognized by various structures and other proteins in the cell. Once the protein is transported to its final destination (e.g., cell nucleus, cell membrane, extracellular space), those signal sequences may be removed. This is another example of posttranslational modification.

A final example (though not the last type of modification), especially common in the production of hormones, is the sequential removal (cleavage) of specific parts of the amino acid sequence that allows regulation of the release of the final hormone protein. Thus, some hormones are known to have pro-hormone sequences, from which the mature hormone (the final version) sequences are produced. Insulin, produced from pro-insulin, is an example.

GENETIC CODE

In order for all cells in the various tissues of the body to function efficiently and communicate effectively, they follow general rules that hold true for all cell types. With regard to DNA biology, several key rules apply, as we've already seen: DNA is made of the nucleotide bases A, T, G, and C, of which A binds with T and G binds with C in the double helix; RNA is composed of the bases U, A, C, and G, which are complementary to the DNA bases A, T, G, and C, respectively; the RNA polymerase machinery transcribes the DNA sequence into a corresponding RNA sequence; and the ribosome translates mRNA to protein. This final process, translating mRNA sequence to amino acid sequence, requires one final rule, known as the genetic code. The **genetic code** is a uniform rule across all cell types in all tissues—that any particular three-letter combination of RNA sequence codes for the same amino acid in all cell types in all tissues. In other words, a specific series of three nucleotide bases (i.e., a *codon*) always contains the same information in all situations. For example, the codon CUU in the mRNA sequence always codes for the amino acid leucine; the codon GCA always codes for the amino acid alanine. Note that all translation begins at the initiation or start codon (AUG) in the mRNA sequence. This three-letter codon always codes for the amino acid methionine. As shown in table 3.1, all of the 64 possible three-letter combinations of the nucleotide bases code for specific amino acids or stop codons. Remarkably, the genetic code is the same for nearly all species on the planet, animal or plant!

Two things are apparent when we look at the genetic code. First, there are three codon sequences

KEY POINT

The genetic code is the uniform rule by which the ribosome converts mRNA codons to specific amino acids. Translation begins with AUG, methionine, and ends when the ribosome reaches any of three possible stop codons (UGA, UAA, UAG).

that fail to code for an amino acid. These are the termination or stop codons mentioned earlier that are recognized by ribosomes as the end point for translation. Second, there are several codons that code for the same amino acid. This makes sense, as there are 64 three-letter combinations of the nucleotide bases but only 20 amino acids used in generating proteins in the cell. For example, the amino acid proline (PRO) is coded for in the mRNA sequence by the following codons: CCU, CCC, CCA, and CCG.

Table 3.1 The Genetic Code

First position	Second position				Third position
	U	C	A	G	
U	UUU Phenylalanine	UCU Serine	UAU Tyrosine	UGU Cysteine	U
	UUC Phenylalanine	UCC Serine	UAC Tyrosine	UGC Cysteine	C
	UUA Leucine	UCA Serine	UAA **Stop**	UGA **Stop**	A
	UUG Leucine	UCG Serine	UAG **Stop**	UGG Tryptophan	G
C	CUU Leucine	CCU Proline	CAU Histidine	CGU Arginine	U
	CUC Leucine	CCC Proline	CAC Histidine	CGC Arginine	C
	CUA Leucine	CCA Proline	CAA Glutamine	CGA Arginine	A
	CUG Leucine	CCG Proline	CAG Glutamine	CGG Arginine	G
A	AUU Isoleucine	ACU Threonine	AAU Asparagine	AGU Serine	U
	AUC Isoleucine	ACC Threonine	AAC Asparagine	AGC Serine	C
	AUA Isoleucine	ACA Threonine	AAA Lysine	AGA Arginine	A
	AUG Methionine **(Start)**	ACG Threonine	AAG Lysine	AGG Arginine	G
G	GUU Valine	GCU Alanine	GAU Aspartic acid	GGU Glycine	U
	GUC Valine	GCC Alanine	GAC Aspartic acid	GGC Glycine	C
	GUA Valine	GCA Alanine	GAA Glutamic acid	GGA Glycine	A
	GUG Valine	GCG Alanine	GAG Glutamic acid	GGG Glycine	G

SUMMARY

DNA is transcribed into mRNA, which is then translated into protein. This process begins in the cell nucleus, where the mRNA molecule is generated. The mRNA molecule undergoes various modifications and moves to the cytoplasm. This modified mRNA is then targeted by the ribosome for the production of the complementary protein, which itself can undergo modifications. All of this takes place under the limitations outlined by the complementary nature of the nucleotide bases (A, G, C, T and U) and the requirements of the genetic code.

KEY TERMS

alternative splicing
codon
gene expression
genetic code
initiation codon
posttranscriptional modification
posttranslational modification
ribosome
RNA polymerase
splicing
start codon
stop codon
termination codon
transcription
translation

REVIEW QUESTIONS

1. Describe the process of transcription. What parts of the DNA sequence of a gene are transcribed into the mRNA? Where does transcription occur?
2. Describe the process of translation. What happens to the mRNA sequence prior to translation? Where does translation occur?
3. What are the signals recognized by the ribosome for starting and stopping the process of translation? How many such stop signals are possible?
4. Describe the genetic code.

CHAPTER

4

MOVING GENETIC MATERIAL TO THE NEXT GENERATION

With the basics of DNA biology in place, chapter 4 moves to the issue of reproduction, or how genetic material moves from generation to generation. In other words, we begin to understand the processes involved in the *inheritance* of DNA sequence across generations. These processes play an important role in the individual genetic variation that an offspring will carry, which is important for an understanding of how genetic factors can influence health and fitness traits.

BASIC REPRODUCTION

Stated quite simply, reproduction is the joining of the male sperm with the female egg (ovum) leading to the union of genetic material from the parents and the development of a new individual, unique from both parents. Our interest in reproduction stems from the fact that genetic material is passed on from parent to offspring in a unique way, and the manner in which that DNA is passed on has consequences for understanding how genetic factors are linked with various phenotypes (as will be explored throughout the text). By understanding the mechanics underlying the combining of parental DNA in a new, unique offspring, we lay the foundation for understanding the complexities of DNA influences on the various phenotypes of interest.

Reproduction is actually the end point of a series of complex processes within the **germ cells** that develop into sperm and egg. Those processes, outlined in detail in the next section, result in each sperm and each egg carrying half of the genetic complement of each individual parent. Moreover, the genetic complement carried by each sperm and egg is randomly generated from the entire genetic complement in each parent, thus providing new DNA sequence, and therefore variation in the traits, of each offspring.

The **sex cells** or **"gametes"** (sperm and egg) are produced by the reproductive organs specific to each sex, namely the testes in males and the ovaries in females. In men, sperm are produced in the testes through a series of steps known as spermatogenesis. Within the testes, spermatogenic germ cells undergo a series of cell divisions and transformations resulting in final sperm formation. Within the ovaries in women, oogonia germ cells undergo a similar transformation, known as oogenesis, resulting in the formation of ova or egg cells. Fertilization occurs when a single sperm fuses with a single egg to form the fertilized egg, or zygote, which can go on to develop into an embryo, a fetus, and eventually a unique individual living outside the mother's womb.

Each sex cell contains only half of the genetic complement required for development of a new organism, with the ova always containing 22 autosomes and one X chromosome, and the sperm always containing 22 autosomes and either an X or Y chromosome. The sperm and egg combine such that the zygote contains the full complement of genetic material: 44 autosomes and two sex chromosomes (XX or XY). Thus, the sperm's DNA content determines the sex of the offspring, depending on which sex chromosome is carried by the sperm that fertilizes the egg. Because mechanisms are in place within the egg to ensure that only one sperm can fertilize it, additional chromosomes from other sperm cells are not a possibility.

SEX CELL PRODUCTION AND MEIOSIS

For the purposes of understanding inheritance and genetic factors, we are more concerned in this text with the building of the sperm and egg cells than we are with what happens after they join. In both men and women, specific germ cells begin the process of sex cell production, which is simply a process of taking one cell and using it as a model to make more cells, and these additional cells are modified to produce sex cells. This process requires two important tasks: DNA replication (copying), thus allowing each produced cell to contain DNA, and cell division.

Before we consider the sex cells, let's first look at the more common example of typical cell division, as occurs in the development of all body cells (with a different process reserved for sex cells). In the case of body cells, known as **somatic cells, cell division** is the process whereby a functioning cell is duplicated, thus producing two "daughter" cells—each a duplicate copy of the original cell. In the case of somatic cells, the daughter cells can be thought of as clones of the original cell: exact replicas. One of the major events leading up to cell division is the process of **DNA replication,** whereby the DNA within the nucleus of the original cell is copied such that one complete copy exists for each of the two daughter cells. Once the genomic DNA is replicated, the process of separating out the duplicated DNA into two daughter cells is known as **mitosis,** and it occurs in nearly all body cells during growth and development.

The exception to this rule is in the production of sex cells, which provide the genetic material for the next generation. The processes of spermatogenesis and oogenesis involve cell divisions that *do not* result in exact replication of DNA material, but rather generate cells containing a slightly different DNA sequence compared to the other body cells. In order to prevent the production of offspring that are exact clones of a parent, the sex cells must have a different DNA sequence compared to either parent's DNA. This process is accomplished through a special cell division known as **meiosis.** The details of meiosis and mitosis are outlined in the Special Focus section.

Meiosis occurs only in the sex cell-forming germ cells of the testes and ovaries. In each case, a single oogonium or spermatogonium undergoes a process of DNA replication and division that results in the formation of four daughter cells, each of which contains only half of the genetic material of the other body cells. Not only is the genetic material halved in each of the sex cells (such that their combination will result in the full complement of DNA), but the genetic material is unique from the other body cells: The DNA sequence has been slightly modified in order to prevent a clone offspring. Thus, meiosis produces four gametes that carry with them a shuffled complement of DNA sequence that, when combined with the opposite-sex gamete during fertilization, can result in a genetically unique offspring. The process of forming DNA sequence that is unique from that of the parents has important consequences for the study of genetic factors, which we will see throughout the text.

KEY POINT

Mitosis is the process of cell division in somatic cells, resulting in exact duplication of a cell's DNA prior to cell division. Meiosis is a unique process of cell division found only in sex cell-forming germ cells, resulting in DNA duplication that is not an exact replication of the original germ cell.

SPECIAL FOCUS

Mitosis Versus Meiosis

Just how do mitosis and meiosis differ? As mentioned briefly in the main text, mitosis is the process by which all but the sex-determining cells go about separating the replicated DNA into the two daughter cells during cell division (figure 4.1). As a somatic, or body, cell begins the process of cell division, DNA is replicated such that the entire complement of DNA is copied exactly. Once the DNA is fully replicated, the cell enters mitosis, which consists of four main

(continued)

SPECIAL FOCUS

(continued)

phases—*prophase, metaphase, anaphase,* and *telophase*—with each phase merging smoothly into the next phase in a seamless process. *Prophase,* the first phase of mitosis, begins when the replicated chromosomes condense into their tightly compacted form. Each individual chromosome, known as a *chromatid,* is paired with its duplicate (e.g., chromosome 1 pairs with the other chromosome 1) by way of a *centromere* that attaches near the middle of each chromatid. Prophase continues as the condensed, paired chromosomes begin to move toward the center of the cell nucleus and the nuclear membrane begins to break up. *Metaphase* occurs when all of the chromosome pairs have aligned in the middle of the cell, their centromeres aligned along the equator or exact center of the cell. At this point, the paired chromatids split at the centromere, with each chromatid separating toward the opposite poles of the cell. This stage is known as *anaphase,* which is the shortest of the phases of mitosis. Following the separation of the chromosomes, a single complete copy of the DNA sequence (a complete genome) is now found at each of the poles, and telophase begins. *Telophase* is the process of forming a new nuclear membrane around each mass of chromosomes at each pole of the cell and the decompaction of the condensed chromatids. This marks the end of mitosis, which is followed closely by cell division or *cytokinesis,* the splitting of the cytoplasm such that two complete cells, exact duplicates of each other, are now formed. This small amount of text does not do justice to either the beauty or complexity of this process, but it provides a foundation for comparing mitosis to meiosis, a different cell division process in sex cells.

Meiosis is similar to mitosis in that it is the separation of replicated DNA among daughter cells. The process differs, though, in that the daughter cells are not clones of the original cell, but rather contain new combinations of the original genomic material to pass along for reproduction (figure 4.2). Moreover, while the general phases of meiosis are similar to those in mitosis, meiosis consists of *two* cell divisions, rather than one, so each phase is distinguished as I or II depending on the specific division (first or second) during which that phase is occurring.

Meiosis begins after the DNA has been replicated in the original cell. This DNA replication event is the same as that occurring in mitosis, so the original cell now contains two complete copies of the genetic material. In *prophase I* (occurring in the first cell division), the replicated chromosomes begin to condense and move to the center of the cell; but in the case of meiosis, all **homologous chromosomes** (e.g., both original chromosomes and their two new duplicate copies) align next to each other along the equator of the cell in a process known as *synapsis.* Thus, four chromosomes are aligned together in a *tetrad.* It is following synapsis and tetrad

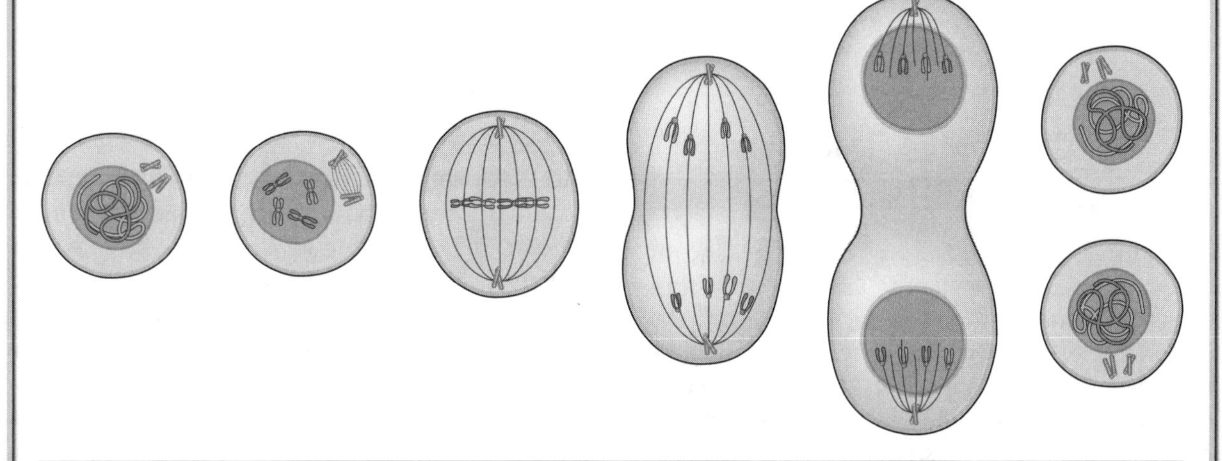

Figure 4.1 The basic steps of mitosis, the process of DNA replication, and cell division for somatic cells.

(continued)

SPECIAL FOCUS

(continued)

formation that the process of **crossover** occurs, in which parts of the chromosomes are exchanged between each other, as discussed in the main text. Following crossover, the genetic material has been randomly mixed among original chromosomes, producing DNA sequences that are different from those of the other body cells. Thus, during *prophase I*, the DNA material of the original cell is modified such that the daughter cells will carry DNA sequences unique from that of the original cell. The cell then undergoes *metaphase I, anaphase I,* and *telophase I,* during which each homologous pair of chromosomes is separated into two daughter cells (now containing new DNA sequence). These two unique daughter cells now undergo *meiosis II.* Meiosis II actually parallels the phases of mitosis, except that it occurs without a round of DNA replication. In other words, the newly formed daughter cells are split, without any copying of the DNA material. The DNA complement in each of the daughter cells is simply separated such that only one-half of the genetic material is found in each of the final four daughter cells present after the end of meiosis II. These final daughter cells are then transformed into the gametes, sperm or egg, each containing half of the genetic material needed for the production of an offspring during reproduction. Neither the sperm nor the egg will carry DNA sequence that exactly matches that of the offspring's parent, ensuring the formation of a unique new individual.

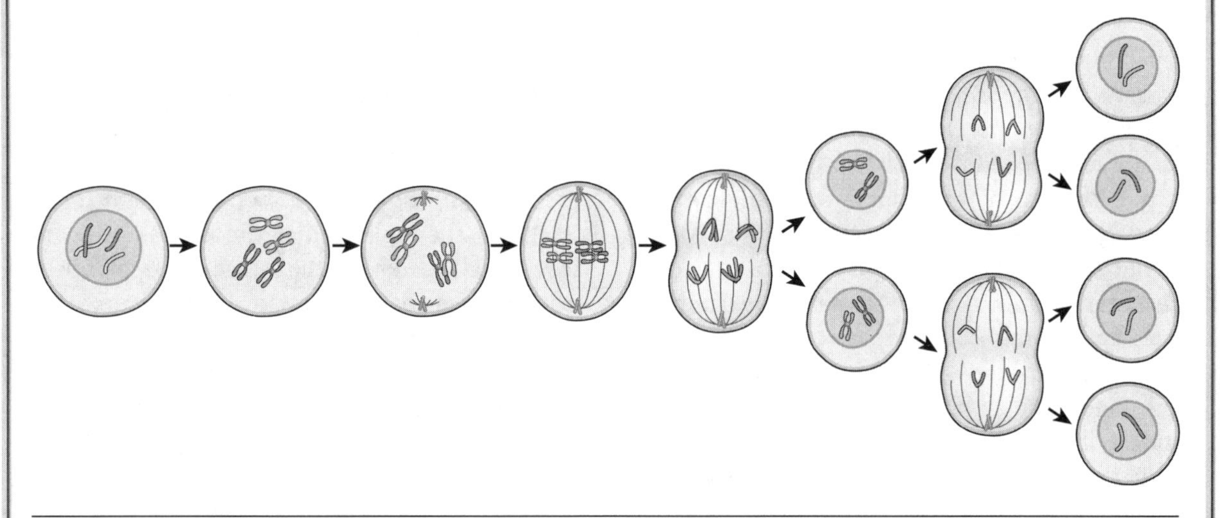

Figure 4.2 The basic steps of meiosis, the process of DNA replication, chromosomal recombination, and cell division for the germ cells of the body.

CHROMOSOMAL RECOMBINATION

A key to generating offspring that are genetically unique from each parent is to "mix up" or shuffle the DNA sequence found in each parent prior to reproduction. That process of mixing the original DNA sequence found in each parent is known as **chromosomal recombination,** which occurs during meiosis (see Special Focus section). Two key processes ensure mixing of DNA in the formation of the sex cells: independent assortment and crossover.

Recall that the end product of meiosis is four gametes, each with 23 chromosomes, half the number required for life. Though each gamete will contain one of each autosome and one sex chromosome, which of the four homologous chromosomes available at the beginning of meiosis moves to a particular gamete is a random process. In other words, because the chromosomes move to different gametes in an independent manner, the existing genetic material is randomly moved to the gametes, the only rule being that each gamete must contain one copy of each autosome plus one sex chromosome. Because there are 46 original chromosomes, from which 23 are selected for each gamete, thousands of different combinations of maternal and paternal chromosomes are possible for any one gamete. This process of random separation of the different chromosomes into the gametes is known as **independent assortment.** This random process ensures the inheritance of unique combinations of each parent's genetic material in the offspring.

> **KEY POINT**
>
> Chromosomal recombination occurs during meiosis to shuffle the DNA sequence prior to distribution to the resulting gametes.This process ensures that the resulting offspring are not exact clones of either parent.

But independent assortment is not the only mechanism for ensuring genetically unique offspring. The second means of ensuring chromosome recombination is known as crossover, whereby parts of homologous chromosomes are exchanged between each other, forming DNA sequences that are unique from those of the original cell. In this process, the four homologous chromosome strands (formed after DNA replication) align in a linear manner during meiosis prophase I, such that each gene on a chromosome is aligned with its homologous gene pair on the other chromosome strand (e.g., the four copies of chromosome 12 align). This is known as a tetrad (see figure 4.3). At this point, a **chiasma** or breakpoint can occur between two of the homologous chromosomes, such that matching sections of those chromosomes can be exchanged. In other words, a section of one chromosome is literally swapped for that same section from the homologous chromosome, resulting in an exchange of DNA sequence between the pair. It is important to recognize that this process does not change the overall structure of the chromosomes; the genes carried and the positions of those genes on the chromosome are exactly the same after the crossover event. Rather, the spellings of these chromosomes will be different from those of the parents' somatic cells, being combinations of the spellings of the original chromosome sequences and reflecting the varying contributions of the original maternal and paternal DNA sequences in the newly formed DNA sequence.

Some details about the crossover process are useful to consider from the perspective of genetic variation. First, only two of the four homologous chromosomes aligned in the tetrad appear to undergo crossover at any one point in time. The two untouched chromosomes may act as templates to ensure exact exchange of genetic material between the crossover pairs, though these "template" chromosomes may later undergo crossover themselves. Second, crossovers can occur in multiple places along a chromosome, though there appear to be specific locations along each chromosome where the breakpoints for crossovers will occur (we'll tackle this in chapter 7). In other words, large blocks of DNA can be exchanged

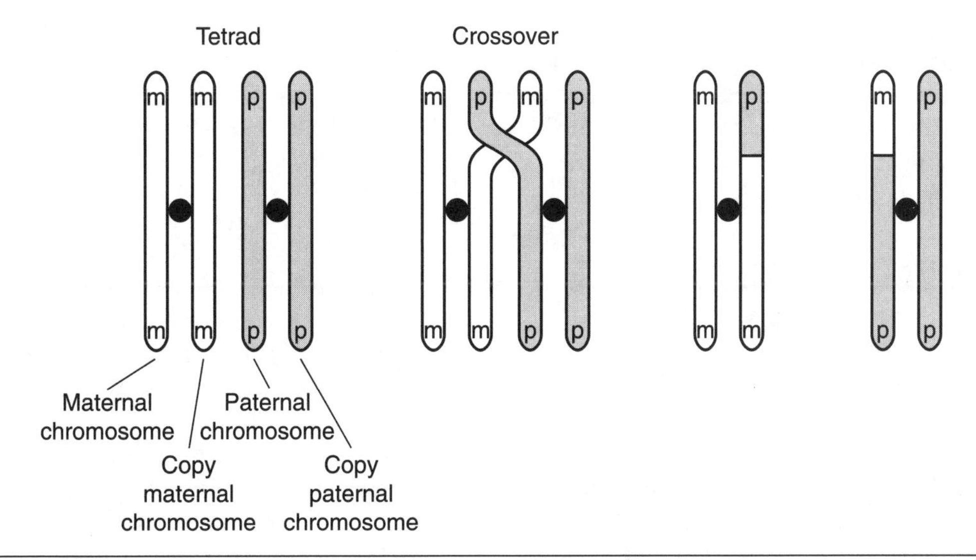

Figure 4.3 The process of crossover, which occurs during prophase I of meiosis. The end result is a new chromosome DNA sequence with the same overall geographical structure as the original chromosome (i.e., same genes and gene locations, but different spellings for those genes).

at various points between the two homologous chromosomes, but the process is not completely random. In the end, the four homologous chromosomes will have DNA sequence different from the original DNA sequence, thereby preventing the reproduction of an exact clone offspring of either parent. The bottom line with the crossover process is that existing DNA sequence is shuffled without the creation of new genetic sequence or chromosome structure.

Some special consideration regarding the sex chromosomes is required at this point. In females, the two X chromosomes undergo all of the processes that we discussed for the autosomes, in a similar manner. The female has two X chromosomes, which are duplicated and can form a tetrad able to undergo crossover similar to the autosomes, so there is no exception in this case. In males, however, having only a single X chromosome and a single Y chromosome prevents the typical pattern of chromosomal recombination. Both the X and Y chromosomes are duplicated in males prior to meiosis I, but that leaves only two of each, preventing the formation of a tetrad as is seen for the male autosomes. Thus, crossover does not occur for the X or Y chromosome in the male. So, female offspring will receive an unmodified X chromosome from the father, paired with a modified X chromosome (i.e., shuffled via recombination) from the mother, while male offspring will receive an unmodified Y chromosome from the father and a modified X chromosome from the mother. The lack of recombination of the Y chromosome means that, barring novel genetic mutations (as discussed in chapter 5), the same Y chromosome DNA sequence is passed along through many, many generations from grandfather to father to son to grandson, and so on, making it useful in the investigation of human migration patterns and other genetics studies. Another consequence of this pattern is that the male's sperm determines the sex of the offspring, not the female's egg.

MENDELIAN INHERITANCE PATTERNS

Because the process of reproduction results in a predictable pattern of DNA sharing between two parents, the offspring of those parents can be predicted to carry certain combinations of genetic information depending on the DNA sequence of the parents. Such inheritance patterns are called **Mendelian inheritance patterns** after Gregor Mendel, an Austrian monk who showed through plant breeding experiments that parental traits are passed on to offspring as discrete "heritable units." Through his extensive breeding of different varieties of pea plants, Mendel showed that specific combinations of traits appeared in the offspring plants in predictable ways, leading to the theory of heritable units, which we now think of as genes. These inheritance patterns also exist in human reproduction, as we will see in this section.

Inheritance of Autosomes

Because each parent contributes one of each of the autosomes to an offspring, the offspring ends up with two copies of each autosome, one from the mother and one from the father. The sperm and egg, each carrying only one copy of each autosome and one sex chromosome, combine such that the zygote offspring has the full DNA complement. Because the sperm and egg combine in such a manner, we can predict the genetic complement of the offspring depending on the genetic "input" of the parents. Let's consider the example of a genetic predisposition to a specific disease. The DNA sequence of the gene in one of the parents is such that the gene spelling results in disease risk (i.e., a "disease gene"). This is not some novel gene not carried by anyone else, but is rather a different spelling of a gene that all people carry; this particular spelling, however, results in disease susceptibility. Because each parent has two copies of each gene, we can predict whether or not an offspring will receive the disease susceptibility gene by knowing the genetic complement of each parent. Suppose that the mother has only one copy of the disease gene (the other copy of the gene has the typical spelling), while the father has two copies of the typical spelling of the gene. Thus, out of four possible genes that can be passed along to the offspring, only one is a disease gene. Thus, any offspring of those parents will have a 50% chance of receiving the disease gene (see figure 4.4, ex. 4). If one parent has two copies of the disease gene, then each offspring is certain to receive one copy of the disease gene (100%). If both parents have one copy of the disease gene each, then each offspring has a 75% chance (two out of four gene copies are disease genes) to inherit the disease gene. Finally, if the father has two disease gene copies and the mother has one disease gene copy, the offspring is certain to receive one copy of the disease gene (from the father) and has a 50% chance of receiving a second disease gene copy from the mother.

Understanding the basic idea of inheritance and the probability of receiving any particular gene copy leads us to the next issue: Is one gene copy more important

KEY POINT

Each parent has two copies of each autosomal gene, one of which will be randomly passed on to an offspring with that offspring carrying two copies, one from each parent. Mendelian inheritance patterns show the probability of an offspring's carrying different combinations of the parental gene copies.

than the other in determining the trait of interest? In fact, one gene copy *may* have more or less influence than the other gene copy in the formation or regulation of a phenotype. In other words, one gene copy may be **dominant** or **recessive** when compared to the other gene copy, which will influence how the phenotype is affected by that gene or its related proteins. These issues are discussed in detail in the next chapter, but the issue of dominance of a gene has implications for predicting the phenotype of an offspring.

With two different gene copies present in an individual, if one copy is *dominant* over the other copy, it will have the largest influence on the phenotype (the other copy is said to be *recessive* to the dominant copy). Think about the disease gene example from our discussion of inheritance patterns. If the disease gene is the dominant gene, only one disease gene copy is needed for the disease phenotype to occur. An example of such a dominant trait is Huntington's disease, a neurological disease that most often results from the presence of just one copy of a mutated HD gene. A second example from the world of sport performance will be detailed in chapter 6. Briefly, a highly successful Olympic champion cross-country skier was found to have a dominant, mutated copy of the erythropoietin (EPO) receptor gene; this made his body more responsive to the red blood cell-promoting actions of EPO. His increased red blood cell and hemoglobin concentrations increased his oxygen-carrying capacity, presumably aiding in his aerobic performance. In this case, having one copy of this dominant gene copy resulted in an altered phenotype. Fortunately in this case, the end result was not disease but rather improved performance!

If the disease gene is recessive, one copy of the disease gene will not result in the disease phenotype, because the typical gene spelling (without disease

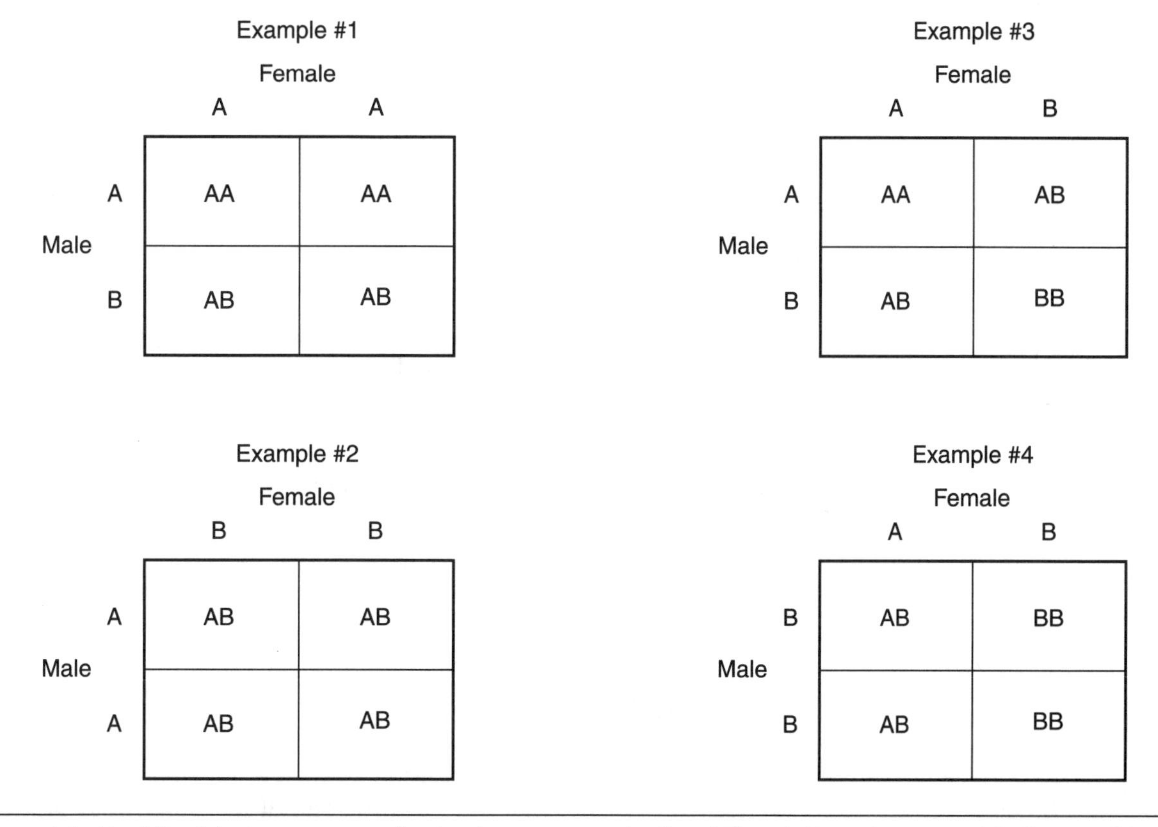

Figure 4.4 Mendelian inheritance patterns, showing the two gene copies (A and B) carried by each parent and the four possible outcomes for passing those genes to offspring. These examples match those described in the text. Notice that when the genetic complement is known for each parent, the probability of an offspring's having a specific genetic complement can be determined.

susceptibility) will prevail in the regulation of the phenotype. Thus, for a recessive gene, both copies of the gene must be of the disease spelling in order for the disease phenotype to develop. An example of a recessive phenotype is cystic fibrosis, a chronic respiratory disease, which develops only when an individual carries two copies of the mutated gene. Examples of dominant and recessive inheritance patterns are shown in figure 4.5.

Not all genes have dominant or recessive actions; some have equal levels of influence, regardless of their disease susceptibility. In this case, we think of the genes as being **additive** in nature, such that each gene copy contributes an equal amount to the phenotype. In this case, if one disease gene copy is present, there is an increased risk of the disease phenotype; if two disease gene copies are present, there is an even greater risk of the disease phenotype (e.g., twice the risk compared to just one disease gene copy).

KEY POINT

A gene can have dominant, recessive, or additive effects on a phenotype. Dominant genes regulate a phenotype when only one copy is present; recessive genes require the presence of two copies; and additive gene copies each contribute equally to the regulation of the phenotype.

Inheritance of Sex Chromosomes

The inheritance of the sex chromosomes requires special attention. Female offspring will receive from each parent one X chromosome, resulting in the full female complement of two X chromosomes, which follows the typical inheritance patterns discussed earlier for the autosomes. In male offspring, however, the pattern is unique. Male offspring are defined by the presence of a Y chromosome, which carries genes unique from all other chromosomes required for male sexual development. Because only the father can pass on a Y chromosome (i.e., the Y is one of the four sex chromosomes available for the offspring from the two parents), male offspring necessarily inherit their father's Y chromosome, which was unchanged during the process of chromosomal recombination (as discussed earlier). The male offspring's X chromosome must come from the mother, which will be unique in DNA sequence from either of the two X chromosomes of the mother because of the process of chromosomal recombination.

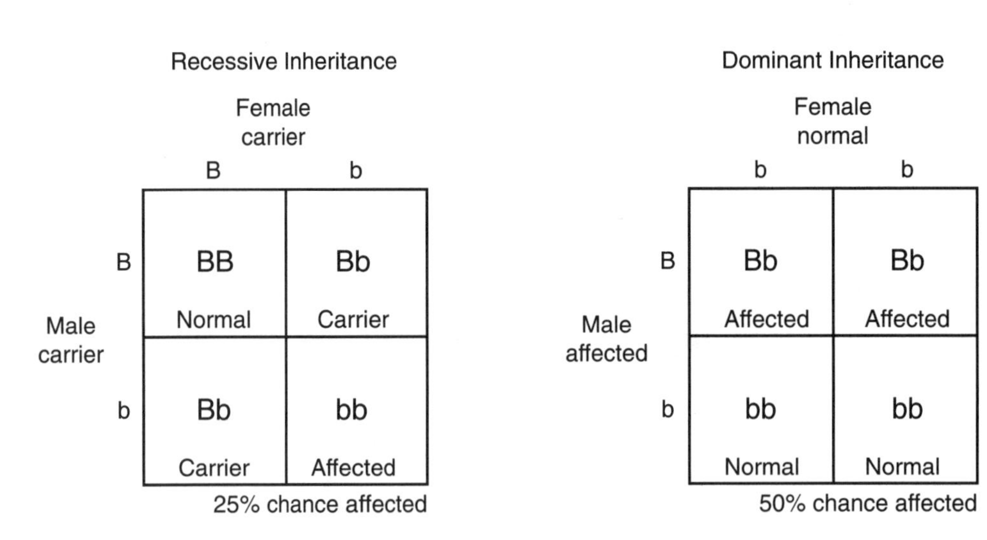

Figure 4.5 Mendelian inheritance patterns with a dominant or recessive trait. A recessive allele must be present in both gene copies for the development of a recessive trait, while only one copy of the dominant allele is required for the development of a dominant trait.

SUMMARY

In this chapter, the basic concepts of reproduction and genetic inheritance were discussed. The key ideas are that offspring are unique from either of their parents because of the process of chromosome recombination during meiosis, or the development of the sex cells. The processes of independent assortment and crossover provide a unique complement of DNA sequence to be passed along to the offspring, thus preventing the reproduction of exact clones of either parent. Because each individual carries two of each autosome, the inheritance patterns of specific genes can be predicted in offspring, including prediction of the presence of a dominant or recessive gene or trait. These basic concepts set the stage for understanding the specifics of genetic variation, which is the topic of the next chapter.

KEY TERMS

additive
cell division
chiasma
chromosomal recombination
crossover
DNA replication
dominant
gametes
germ cells
homologous chromosome
independent assortment
meiosis
Mendelian inheritance patterns
mitosis
recessive
sex cells
somatic cells

REVIEW QUESTIONS

1. What chromosomes are contained within a somatic cell? What chromosomes are contained within the daughter cells after mitosis and cell division of that somatic cell?
2. What chromosomes are contained within a germ cell? What chromosomes are contained within a gamete after meiosis and cell division of that germ cell?
3. Describe the two processes during meiosis that ensure that unique genetic information will be passed along to offspring.
4. How are the processes of chromosome recombination and inheritance different in males compared to females?
5. As was done in figure 4.5, determine the chance that an offspring will carry two copies of a recessive disease gene when one parent is affected and one parent is unaffected but carries the disease gene.

CHAPTER

5

HERITABILITY AND THE BASICS OF GENETIC VARIATION

As discussed in the introduction to this book, there is plenty of evidence to suggest that genetic factors play a role in the variability that we see in various health- and fitness-related phenotypes among individuals. In chapter 5, we identify genetic factors as DNA sequence variation, or slight differences in DNA sequence that occur across all individuals (with the exception of identical twins). While we as humans share over 99.9% of our DNA sequence in common, the 0.1% that differs is significant enough to make each of us unique from one another. While we all share the same chromosome structure and the same complement of genes, the spellings of those genes are slightly different among individuals; this is the essence of what we refer to as genetic factors.

FAMILIAL RESEMBLANCE: GENES AND ENVIRONMENT

As you look in the mirror, who do you see besides yourself? Often, we recognize physical features that we share with our extended family members, and such familiarity can be seen across multiple generations. Although even unrelated individuals share similar DNA sequences (99.9% similar), families share an even greater fraction of their genome sequence and carry many more of the same DNA sequence variations compared to unrelated individuals. It is important to recognize also that families share more than just very similar genetic sequence; families also often share the same or very similar environments. In other words, families, especially close-knit or nuclear family members, often share diet patterns, physical activity patterns, leisure-time patterns, educational activities, and so on. So both genetic variation and environmental factors are important in determining trait resemblance among family members. This combination of factors poses a problem for genetics researchers, as variation in a trait is likely to be caused by a mix of various genetic and environmental factors, and distinguishing among them can be quite challenging.

The first step in determining the importance of genetic variation in any trait is to identify **familial aggregation,** or resemblance across family members. To do this, researchers recruit siblings, sibling twin pairs, a mother and her children, or other family member combinations, and perform **correlation** analyses in order to determine how closely the individuals within the various combinations resemble one another for a particular trait. If there is a strong correlation among family members, meaning that members of a family have similar measures of a phenotype, then there is familial aggregation. What can't be determined from a simple correlation among family members is whether such phenotype similarity is due to genetic or environmental factors. Figure 5.1 shows an example of familial aggregation from the HERITAGE Family Study.

Recall from chapter 1 that the members of each family recruited for the study performed the same 20-week exercise training program and that considerable variability was observed for the response of $\dot{V}O_2$max among the subjects. In figure 5.1, we see that the variability among subjects could partially be

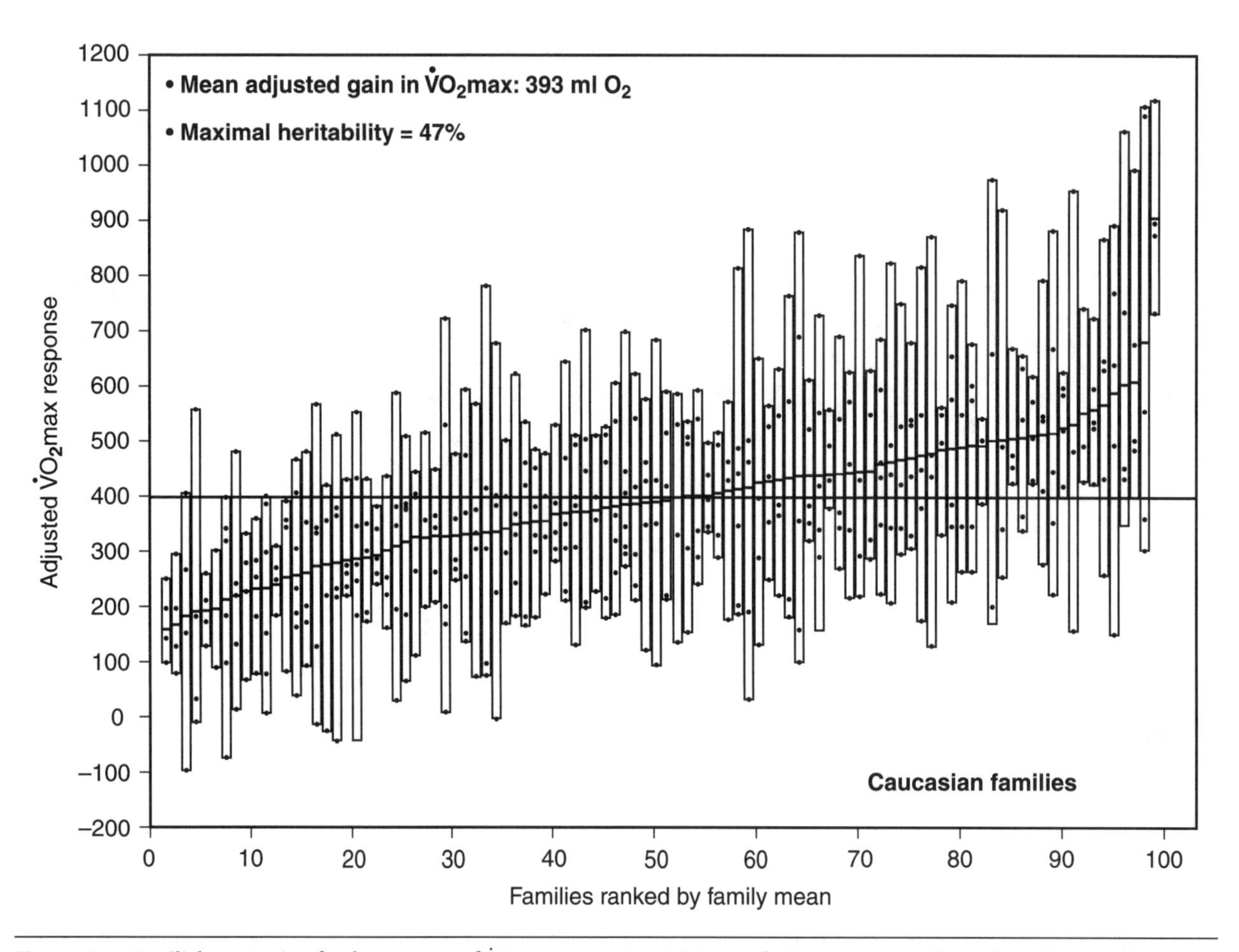

Figure 5.1 Familial aggregation for the response of $\dot{V}O_2$max to exercise training in the HERITAGE Family Study. Each vertical bar represents a single family, with each dot within a bar showing the change in $\dot{V}O_2$max value for an individual family member. The horizontal line near the middle of each bar represents each family's average $\dot{V}O_2$max response.

Reprinted from C. Bouchard et al., 1999, "Familial aggregation of $\dot{V}O_2$max response to exercise training: results from the HERITAGE Family Study," *J. Appl. Physiol.* 87(3): 1003-1008. Used with permission.

predicted from knowledge of the specific family of that individual. Notice that some families have generally lower responses to exercise training and others have higher responses to exercise training than other families. Notice also that even within families there can be considerable variability, meaning that more than genetic factors is important for determining the response of $\dot{V}O_2$max to exercise training.

HERITABILITY AS A MEASURE OF GENETIC CONTRIBUTION

Once familial aggregation or resemblance has been shown for a phenotype, researchers seek to determine the extent to which both genetic and environmental factors influence that trait of interest. **Heritability** is a quantitative estimate of the contribution of genetic factors to a trait.

In order to estimate the fraction of trait variability that is determined by genetic factors, specific comparisons across family members must be made. Because the extent of DNA sharing can be predicted for any two family members (see chapter 4), we can predict the genetic sequence similarity for any two related individuals. For a trait with a strong genetic influence, individuals with less shared genetic variation (e.g., second cousins) would be expected to be less similar for that particular trait than identical twins, who share 100% of their DNA sequence. Various combinations of this approach can be used to estimate the fraction of familial aggregation that is due to genetic versus environmental factors.

One of the most common methods of deciphering genetic versus environmental influences in familial aggregation is the use of twin pairs. There are two types of twin pairs: monozygotic and dizygotic. **Monozygotic twins** (MZ twins) share 100% of their

DNA sequence and are formed when a single early-stage zygote splits into two zygotes very shortly after conception. Thus MZ twins, also known as **identical twins,** share the same DNA sequence, as the cells of both twins originated from the same fertilized egg. **Dizygotic twin** (DZ twin) pairs, on the other hand, share only ~50% of their DNA sequence and are very similar to any other sibling pair. Dizygotic twins, also known as **fraternal twins,** are formed from two distinct fertilized eggs, both egg cells being present during a single menstrual cycle in a woman. During reproduction, if two egg cells are present and are fertilized by two different sperm, two individuals can form, each with a distinct genetic sequence. Dizygotic twins differ from other siblings, however, in that they shared the same maternal environment in the womb, which will not be the case for siblings born in different pregnancies.

With this understanding of MZ and DZ twins, we can now discuss the most common method of estimating heritability for a trait. In this method, we recruit MZ twins and DZ twins and measure our phenotype of interest in both twins of a pair, and in many such twin pairs from different families. We then test to see how similar the phenotype measurements are across the various twin pairs; this is a *correlation analysis* (often called an "intraclass" correlation). If we assume that all of the twin pairs were raised together in the same family environment, we would expect the twins to show familial resemblance for our trait of interest. But, because MZ and DZ twins share a different fraction of their DNA sequence variation, namely 100% for MZ twins and ~50% for DZ twins, we can estimate the extent to which genetic or environmental factors are contributing to that similarity. In other words, for a trait of strong genetic influence, we would expect that identical, MZ, twins would be more similar (i.e., show a stronger correlation) than would DZ twins, who share less DNA sequence variation in common. Through a comparison of the correlation values for the two types of twins, an estimate of genetic contribution, or *heritability,* can be calculated. An example is presented in figure 5.2, where MZ twin pairs are shown to have a correlation value (r) of 0.85 for a hypothetical phenotype while DZ twin pairs have a correlation of only 0.50. Because the familial similarity is stronger in the identical twins, we predict that genetic factors are important for this trait. All this is *not* to say that environmental factors are not important; in fact, the fraction of trait variation left over after an estimation of heritability is assumed to be due to environmental factors, as we'll see in a moment.

KEY POINT

Monozygotic twins share 100% of their DNA sequence, while dizygotic twins share only ~50% of their DNA sequence, as with any sibling pair.

The calculation for heritability in a twin study such as that shown in figure 5.2 is fairly straightforward. Heritability (denoted as h^2, H^2, or H) is simply the difference between the intraclass correlation values

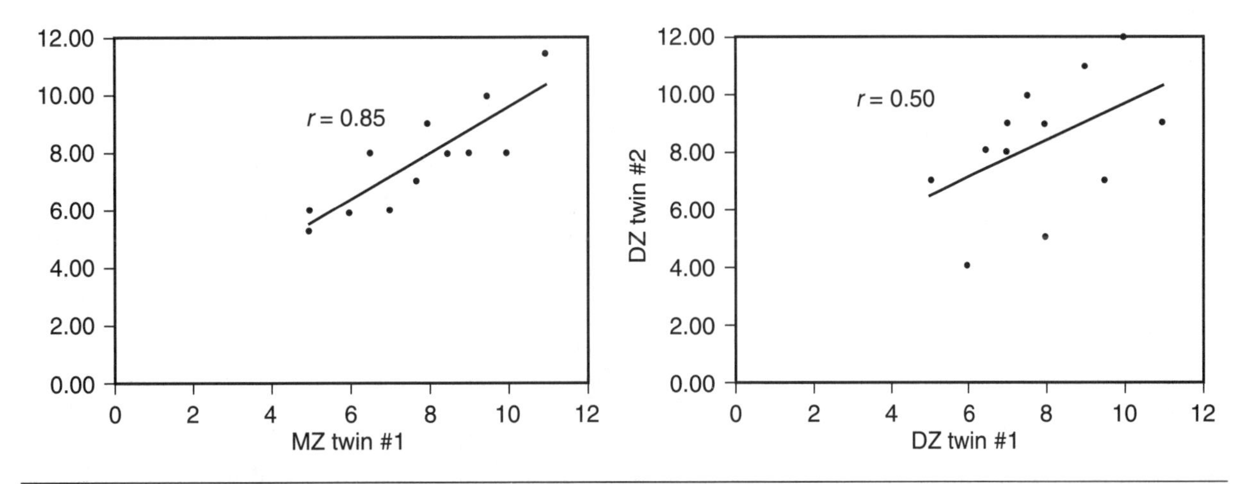

Figure 5.2 An example of a heritability analysis performed for monozygotic and dizygotic twin pairs. Each graph shows the correlations between the twins in the twin pairs, the graph on the left for MZ twins and the graph on the right for DZ twins. The phenotype measure for Twin #1 is graphed with the measurement value for Twin #2 in a scatterplot graph, and the linear regression line and correlation value r are determined for each graph.

(r) of the MZ twin pairs versus the DZ twin pairs. Shown mathematically, the equation is as follows:

$$h^2 = 2(r_{MZ} - r_{DZ})$$

where

h^2 is heritability,

r_{MZ} is the correlation value for MZ (identical) twin pairs,

r_{DZ} is the correlation value for DZ (fraternal) twin pairs, and

h^2 is equal to twice the difference between the MZ and DZ correlation values.

From the example shown in figure 5.2, we would calculate $h^2 = 2(0.85 - 0.50) = 0.7$, or 70% heritability. Alternative equations are also available, such as the following:

$$h^2 = (r_{MZ} - r_{DZ}) / (1 - r_{DZ}).$$

We can think of the total variation of a trait as being 100% (all possible variability of the trait or phenotype among different individuals), to which genetic and environmental factors each contribute some part. We can visualize this in the form of a pie chart. As shown in figure 5.3, the fraction of trait variation assumed to be due to genetic factors (i.e., the heritability or h^2 value) is balanced by environmental factors, such that the two values are assumed to add up to 100% of trait variability. Thus, you will often see h^2 values reported as percentages (e.g., 35%), which can be interpreted to mean that genetic factors are contributing that percent to the total variation measured in the phenotype. Mathematically, we can describe the relationship as follows:

$$V_T = V_G + V_E$$

where

V_T is the total trait variance,

V_G is the variance explained by genetic factors, and

V_E is the variance explained by environmental factors.

From this equation, heritability can be represented mathematically as

$$h^2 = V_G / V_T$$

where heritability (h^2) equals that fraction of the total variation in the phenotype (V_T) explained by genetic factors (V_G).

You've probably noticed that I've used words like "estimated," "assumed," and "predicted" in this section. The estimate of heritability is just that: an educated guess about the extent to which genetic factors

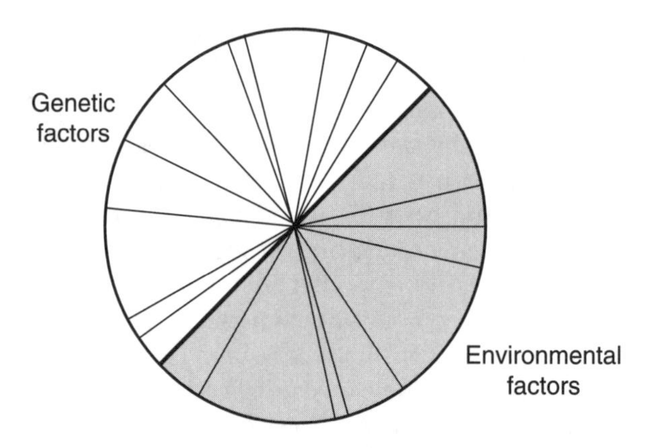

Figure 5.3 Genetic and environmental factors contributing to a trait. In this example, genetic factors and environmental factors each contribute about 50% to the total variance of the trait. Within each of those categories, however, multiple factors are contributing varying amounts. In the case of environmental factors, this means that multiple environmental factors are each contributing to the variation in the trait; the same is true for genetic factors.

influence a trait. Because actual genetic sequence variation in DNA is not being measured in these studies (rather, the analyses are correlation analyses), there is no way to perform a true cause-and-effect analysis. Thus, if you look in the literature for estimates of heritability for a phenotype, you will often find large ranges of heritability values, which is reflective of the inexact nature of this measurement.

KEY POINT 

Heritability is a quantitative estimate of the contribution of genetic factors to a particular phenotype, typically presented as a percentage (0-100%).

GENETIC VARIATION

Once familial aggregation and heritability studies are performed and genetic factors are shown to influence a trait of interest, the next task for researchers is to determine which genetic factors are important for that trait. This is an extremely challenging task! For simple, Mendelian diseases, often a single letter of the 3.1 billion letters in the human genome must be identified. And for complex traits, such as $\dot{V}O_2max$, body fat mass, and cognitive function, more than one gene variant is likely to contribute to the trait, with the potential for various DNA sequence variations in several genes to be important. The technologies required to perform such work have improved dramatically in the last 10 to 20 years, especially with

SPECIAL FOCUS
Twin and Family Studies

From the time of the ancient Greeks, the similarity of twins has been of scientific interest. How is it that two individuals can be identical? Perhaps of more interest has been how two or more siblings could be born of the same pregnancy but not be identical (i.e., fraternal twins). We know today that all family members share very similar DNA sequence and that it is this shared DNA sequence that contributes to shared physical traits and disease predispositions. From the work of Mendel and others, we can estimate closely the extent to which different family members share DNA sequence variation, thus allowing comparisons and estimates of genetic contribution. As J.B.S. Haldane is quoted as saying, "Would I lay down my life to save my brother? No, but I would to save two brothers or eight cousins" (brothers share 1/2 of their DNA sequence, while first cousins share 1/8 of their DNA sequence).

But families also share more than just DNA sequence; they often share unique environmental factors as well, such as diet, physical activity, and other behavioral and cultural variables. The sharing of environmental factors is more variable than DNA sequence sharing, but the correlation is typically high, especially among siblings in a nuclear family. Thus families, with shared DNA sequence and often shared environment, provide an excellent means of estimating the contribution of genetic and environmental factors to traits of interest.

As discussed in the main text, comparisons of identical (or MZ) and fraternal (or DZ) twins can be used to estimate the heritability of traits with familial aggregation or resemblance. MZ twins share 100% of their DNA sequence, while DZ twins share only 50% of their DNA sequence; both types of twin pairs, if raised together in the same family, are likely to share environmental factors (including the maternal environment in the womb). If the similarity of a phenotype in MZ twins is considerably greater than that observed for DZ twins, the indication is that genetic factors play an important role in the trait. Conversely, when the correlation values are similar between MZ and DZ twin pairs, this suggests that differences in genetic factors play a less important role than environment. A variation on the typical twin study is to study twins (especially MZ twins) who have been raised apart, in different environments. Such comparisons provide important insight into the role of genetic versus environmental factors.

But family studies can extend beyond the use of twin pairs. Combinations of family members of different genetic similarity can be equally instructive as to the extent of the contributions of genetic and environmental factors to a phenotype. Consider the example of studying parents plus offspring who live together. The two parents share similar environmental factors but do not share DNA sequence, while the offspring share environmental factors and DNA sequence with both parents. So, if the correlation of a trait between parents is similar to that observed in the sibling offspring, the indication is that environmental factors are more important than genetic factors. Conversely, if little correlation is seen between parents (who share only their environment) but considerable correlation is seen among sibling offspring, then genetic factors are likely playing a larger role in the phenotype. Studies of adopted children who were raised in an environment different from that of the biological parents and other family members can also provide insight into the importance of genetic versus environmental factors in a trait.

The examples just discussed rely simply on correlation data (*r*-values) obtained between various combinations of family members, with fairly straightforward mathematics used to estimate genetic and environmental contributions. But, while these analyses are straightforward to perform, they provide fairly rough estimates for the contributions of genetic and environmental factors. More advanced analyses of families are possible, however, including designs that involve the use of genetic data (i.e., actual DNA sequence variation data). These studies, including the transmission disequilibrium test (TDT) and the sibship disequilibrium test (SDT), among others, are much more statistically sophisticated, allowing more complex analyses of genetic and environmental factors. In fact, these advanced methods can be used not only to demonstrate an important genetic contribution to a trait, but also to identify genomic regions that might contain specific genes important to that trait. The specifics of such methods are beyond the scope of this text, falling under the umbrella of population genetics or statistical genetics.

the complete sequencing of the human genome. Even with these advancements, however, scientists expect to work most of this century before completely deciphering the genetic and environmental factors underlying most complex traits. These issues will be discussed more thoroughly in chapter 6.

Just what do we mean by *genetic factors,* also called **genetic variation**? Most often, the term genetic factors is used to describe slight variation in DNA sequence, such as letter differences in the DNA sequence that can influence the information in a gene or gene region. In other words, while all humans share the same 20,000 or so genes, the spelling of those genes can differ slightly, resulting in the slight phenotype differences we see among all humans. Of our 3.1 billion DNA letters, it is estimated that there is a sequence variation every 100 to 1000 letters, or that individuals all across the world differ by no more than 1% of their DNA sequence and possibly much less. So, while all of us humans share at least 99% of our DNA sequence, that 1% contains many possible letter differences that can account for the individual variability we see in all traits.

Any position in the genome where more than one nucleotide base is found among a population of individuals is known as a **mutation** or **polymorphism** (these terms are distinguished farther on), and each variant letter found at a mutation or polymorphism is known as an **allele.** In the vast majority of cases, two different alleles are found at any polymorphism in the genome (e.g., the nucleotides A and G [alleles] are both possible at position 15 in exon 2 of gene ABC). Across a population, the frequency of each allele can be measured; these measurements are often called the common and rare **allele frequencies,** respectively, depending on which allele is carried more frequently by individuals in the population.

Because most genes come in pairs (the only exception being X and Y chromosome genes in men), each person has a pair of alleles within his or her genome for any polymorphism, one coming from the mother and one from the father. This combination of alleles at any variant site is known as a **genotype.** Thus, a person can have two copies of the same allele, or a copy of each of the variant alleles. For example, at a polymorphism with the nucleotides C and T found in the population, any individual will have one of the following three genotypes: C/C, C/T, and T/T. The term **homozygote** is used to describe an individual carrying two copies of the same allele (e.g., C/C or T/T), while **heterozygote** is used to describe an individual carrying two different alleles (e.g., C/T).

Two terms are important to distinguish here: mutation and polymorphism. Any variant site in the human genome can be called a mutation. In other words, if one person out of the billions of people in the world has a C allele at a certain position in a gene when all other persons have a T allele, this C allele can be thought of as a mutation. All positions of the genome that show nucleotide variability in even one person in the world can be thought of as mutations. In practical terms, however, the frequency of the rare allele found at any particular DNA nucleotide position affects the terminology used to describe it. As in the example just given, the term *mutation* is often used to refer to alleles with very rare frequency, for example a rare allele found in less than 1% of the world's population. Common sequence variants, on the other hand, or those in which the rare allele is found in more than 1% of the population, are called *polymorphisms.* While rare mutations are often the subject of study for rare genetic diseases (e.g., muscular dystrophy, cystic fibrosis), the more common polymorphic genetic variants are the focus of personalized medicine and common-disease researchers and clinicians (discussed in chapters 6 and 12). This text focuses primarily on the influence of polymorphisms on physical traits and disease phenotypes.

TYPES OF GENETIC VARIATION

The most common type of DNA sequence variant is the **single-nucleotide polymorphism,** or **SNP** (pronounced "snip"). As described in the previous sections, 10 to 15 million SNPs are estimated to exist in the human genome across the world's populations. A SNP is simply a single nucleotide position within the DNA that has a variant allele in greater than 1% of the population. The C- and T-allele example from the previous section represents a SNP, with genotypes C/C, C/T, and T/T possible for any individual, assuming that both the C and T alleles are found in more than 1% of the population.

Not all genetic variants are SNPs, though SNPs are the most common sequence variations in the genome and the most frequently studied in human genetics research. As shown in figure 5.4, other types of DNA sequence variants exist as well. **Insertion/deletion polymorphisms** are the presence or absence of a stretch of specific DNA nucleotides at a certain position in the genome. The insertion allele (often denoted as **I allele**) is the presence of this stretch of DNA sequence, while the deletion allele (or **D allele**) is the absence of the sequence. Similar to the situation with SNPs, an individual will carry two insertion/deletion alleles, one for each gene copy, and can be homozygous for either the I (I/I) or D alleles (D/D), or heterozygous, carrying one of each allele (I/D). The length of the insertion/deletion allele can be as small as one nucleotide to as large as several hundred bases. The location of an insertion or deletion can have either little to no effect on gene function or a profound effect on gene function, depending on the exact location of the insertion/deletion position in a gene. We'll discuss the influences of polymorphisms on gene function in the next section. A very commonly studied insertion/deletion polymorphism (or "indel," as it is sometimes abbreviated) exists in intron 16 of the angiotensin-converting enzyme or ACE gene. The insertion in the ACE gene is 287 nucleotides in length. We will discuss the ACE gene and its insertion/deletion polymorphism in depth in chapter 11.

Another type of genetic variation is known as a **repeat polymorphism,** also known as a **microsatellite repeat.** These polymorphisms are repetitive stretches of short DNA sequences (e.g., GGC, CA), and the number of sequence repeats differs among individuals. In the case of a repeat polymorphism, the number of repeats present in a stretch of DNA is used to identify the allele, in effect making the length of the repeat (or the number of times the sequence is repeated) the allele. As seen in figure 5.4, the repeat polymorphism shows the two alleles as three repeats and five repeats, so the individual would be heterozygous at this polymorphism. For example, in the androgen receptor (AR) gene, a repeat polymorphism is found in exon 1. This polymorphism contains a series of repeats of the three-nucleotide sequence CAG, which code for the amino acid glutamine (GLN). People can have as few as nine to as many as 30+ repeats of the CAG sequence; thus the length of the glutamine amino acid chain in the AR protein is variable. In fact, very long lengths of this CAG repeat sequence (e.g., >40 CAG repeats) have been linked with an AR disease known as spinobulbar muscular atrophy or Kennedy disease. Studying these repeat polymorphisms presents a major challenge in that there are many combinations of allele length to study, making it difficult to group people. Often, general categories of repeat length are used to group people for statistical purposes.

Single-Nucleotide Polymorphism (i.e., SNP)
TGG ACC TAG GGC AGT CAG CTG
TGG ACC TAG GGC ATT CAG CTG

Alleles: G and T.

Possible genotypes: G/G, G/T, and T/T.

If the alleles above were found in the same individual, the genotype would be G/T.

Insertion/Deletion Polymorphism
TGG ACC TAG GGC AGT CAG CTG
TGG ACC TAG GGC - - - CAG CTG

Alleles: I and D.

Possible genotypes: I/I, I/D, and D/D.

If the alleles above were found in the same individual, the genotype would be I/D.

Repeat Polymorphism or Microsatellite Repeat
TGG ACC TAG GGC GGC GGC AGT CAG CTG
TGG ACC TAG GGC GGC GGC GGC GGC AGT CAG CTG

Alleles: the number of repeats of GGC present in each sequence, $(GGC)_n$. In this example, the alleles are 3 repeats and 5 repeats.

Possible genotypes: many possible genotypes, reflecting the many possible numbers of repeats.

If the alleles above were found in the same individual, the genotype would be 3 repeats/5 repeats.

Figure 5.4 Examples of different types of DNA sequence variation. Each sequence represents the same DNA sequence in two different chromosomes, perhaps in two different individuals or perhaps on the two chromosome copies within a single individual. Only one strand of the DNA double helix is shown. The alleles and genotypes are distinguished for each polymorphism.

KEY POINT

Genetic variation is any position in the DNA where different nucleotides are found in different individuals in a population. Many kinds of such sequence variation exist, including single-nucleotide, insertion/deletion, and repeat polymorphisms. Such variations are known as mutations when they are seen only very rarely in a population.

Both insertion/deletion and repeat polymorphisms that are located in the coding regions of a gene (i.e., exons) have the potential to drastically alter amino acid sequence. For insertion/deletion polymorphisms, the inclusion (or absence) of a segment of DNA could add or subtract several amino acids from a protein sequence. Likewise, the number of repeats in a repeat polymorphism can drastically affect the number of amino acids present in a protein, as seen with the AR gene's CAG repeat just described. But both of these scenarios are incomplete, in that they suggest otherwise normal protein sequence with the addition or subtraction of a few amino acids. Such a relatively small influence on a protein is possible only if the size of the insertion/deletion or the number of repeats is *divisible by three* (i.e., the size of a triplet codon) and it occurs in a codon-codon boundary. In other words, if the length of the insertion/deletion sequence is not some multiple of three nucleotides, less than a complete codon will be present in the sequence, which will dramatically alter the rest of the coding sequence. This is known as a **frameshift mutation** in the DNA sequence, and an example is shown in figure 5.5. Notice how the insertion of a

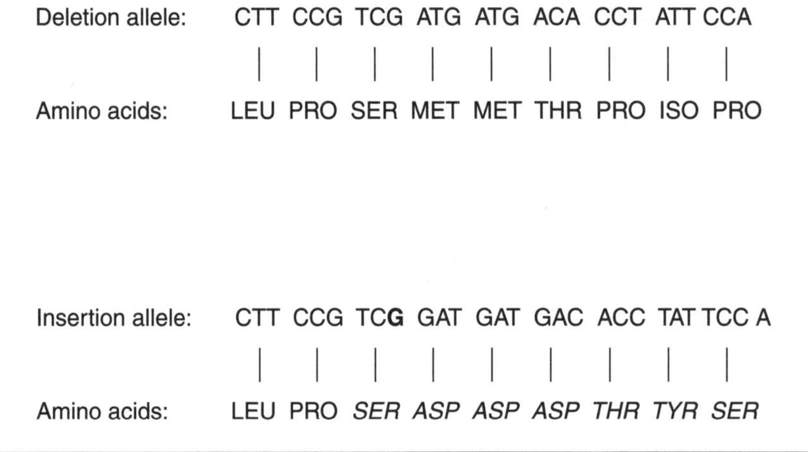

Figure 5.5 An example of a frameshift mutation, in which a single nucleotide insertion is found within the coding region of a gene; notice that the codons within the open reading frame are completely changed downstream of the insertion allele, resulting in new amino acids. The DNA sequence and amino acid sequence are shown in each example.

SPECIAL FOCUS

Labeling Polymorphisms

We've seen how to label a genotype for SNPs (e.g., C/C, C/T, and T/T), but what about the label used to describe the polymorphism itself? The typical polymorphism label contains two important pieces of information: the two alleles possible at the variant site and the physical location of the alleles within a gene. The label does not include the name of the gene, so the polymorphism must be identified as existing in or near a gene in order for researchers to know its location in the genome. For example, a typical label to describe a SNP might be T11C, which describes the SNP as having alleles T and C at position 11. We need more information to place this polymorphism somewhere in the genome, so in describing the polymorphism we would have to include other information, such as that the T11C SNP is located in exon 2 of the ABC gene. For a polymorphism in the promoter of the ABC gene, typical labels will include a minus (–) sign to indicate that the SNP is upstream of the beginning of the first exon (e.g., A–568G). Figure 5.6 shows examples of some typical polymorphism labels.

(continued)

SPECIAL FOCUS

(continued)

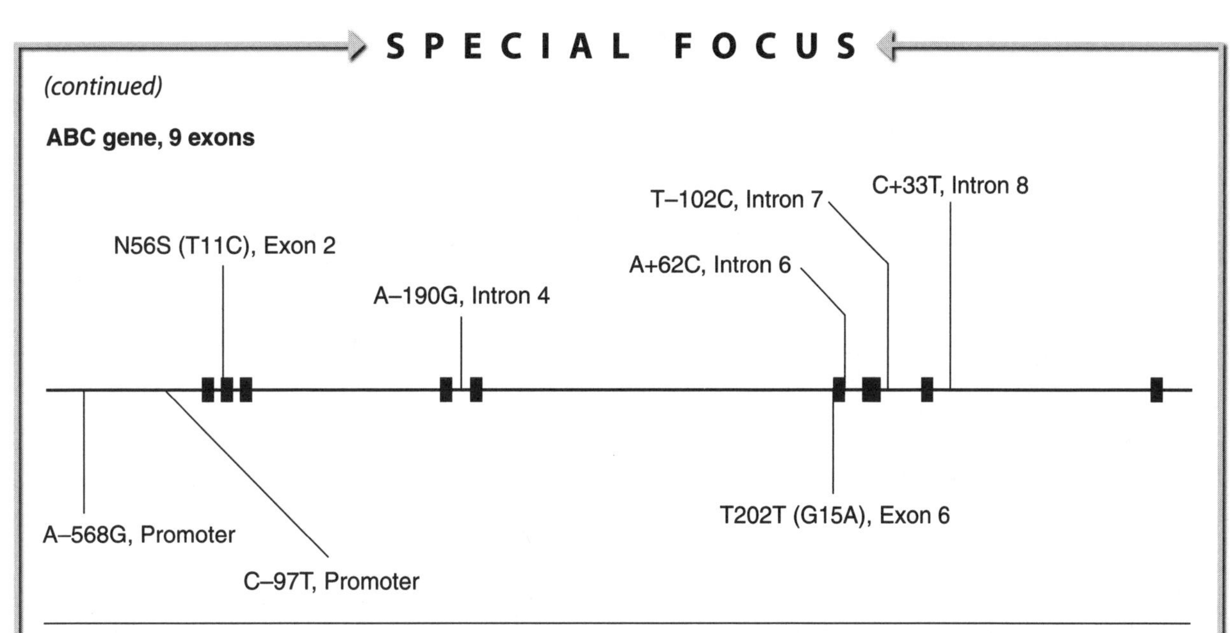

Figure 5.6 An example of polymorphism labeling within a gene. Notice that promoter SNPs are distinguished by a minus sign, but minus signs are also used occasionally for intronic sequences to show that they are upstream of an exon. Coding region SNPs are typically shown with the amino acids in the label (e.g., N56S, T202T), though the nucleotide label can also be used.

In the case of polymorphisms that change an amino acid, very often the two possible amino acids will be shown in the label, along with their codon location, rather than the specific nucleotide alleles and the location of those alleles. This is to emphasize the fact that the protein is different depending on which allele is present, which is important when one is trying to understand the potential importance of a polymorphism in gene function. In this case, either the three-letter or the single-letter abbreviation for the amino acids is used in the label. For example, the label LYS3ARG (or K3R) could be used to label a polymorphism in the third codon of an exon that results in the amino acids lysine or arginine. The true underlying polymorphism label (i.e., DNA sequence) would be different, for example, A8G, with alleles A and G at position 8 in the exon (e.g., codons AAA and AGA, respectively).

Because the letters G, A, C, and T can be used as single-letter abbreviations for both amino acids and nucleotides, some care must be taken to ensure that the label accurately reflects either protein or DNA sequence. When both amino acids have either G, A, C, or T as their single-letter abbreviation, using the three-letter amino acid abbreviations for the label is recommended. These abbreviations are shown completely in table 2.1 in chapter 2.

Polymorphism labeling is, however, imperfect, and unfortunately no standard system of nomenclature has been developed for labeling; this has at times caused confusion for researchers attempting to replicate another group's work (though some groups are working to develop standards to reduce complexity and confusion). The primary problem is that the number used to indicate location within a polymorphism label is up to the researchers who first identify the polymorphism. For example, for the A–568G label discussed earlier, is that SNP located 568 nucleotides upstream of the start of the first exon or upstream of the initiation codon (which may not be at the start of the first exon)? In some cases, the label number is not specific to an exon or intron, but simply reflects the location of the SNP downstream of either the start of the first exon or the initiation codon. The bottom line is that one will often need to read the original source paper in order to get a clear picture of the exact location of the polymorphism and the rationale used for a particular label.

For insertion/deletion and repeat polymorphisms, concise labels such as those used for SNPs are not typical, because more information is needed to fully describe these polymorphisms. In the case of insertion/deletion polymorphisms, we need to know the location where the insertion sequence will be present, and also the length of that insertion. If the insertion falls within a coding region, does it

(continued)

SPECIAL FOCUS

(continued)

alter the amino acid sequence or cause a frameshift? Similarly for repeat polymorphisms, the letters of the repeated sequence, the location of the repeats, and the typical number of repeats (i.e., alleles) all require discussion, in addition to a description of any influence on amino acid sequence for coding region polymorphisms. These polymorphisms require fuller descriptions than are needed for SNPs, and the first papers describing the polymorphism should contain such details. Because these types of polymorphisms are less common than SNPs, however, few are likely to exist in any one gene. Thus, these polymorphisms often become known in the literature simply as the "I/D polymorphism in the ABC gene" or the "CA repeat in the XYZ gene," and the details are left to the research papers that first presented the polymorphism.

single nucleotide dramatically alters the amino acids downstream of that insertion allele. A frameshift would also occur with an insertion allele divisible by three if it occurred within an existing codon (rather than between codons). Whenever a frameshift allele is present in a polymorphism, the likelihood exists for a major functional change for the protein, very often completely disrupting protein function. Thus, frameshift mutations are most often seen in genetic diseases, having been observed in cystic fibrosis, familial hypercholesterolemia, and other rare disorders.

HOW GENETIC VARIATION CAN INFLUENCE PHYSICAL TRAITS

What does it mean to have a genetic influence on a trait or disease of interest? Ultimately, it means that the alleles within a specific gene variant influence gene function in different ways. In other words, the different nucleotide sequences can influence how a gene is turned on or how a coded protein functions. These changes in the amount or function of the gene's protein can then affect the physiology of the cell, tissue, or organ, resulting in measurable phenotype differences at the whole-body level.

This is the idea of a "functional" allele, or an allele within a polymorphism that affects a gene or its protein differently from the other allele at that site. A simple example of this is a SNP within the coding region of a gene. If the SNP lies in an exon, then it also likely lies within a codon within the exon that codes for a specific amino acid. As shown in figure 5.7, if the two codons corresponding to the two different alleles result in different amino acids, then the amino acid sequence of the protein is going to differ depending on which allele is present. Because each amino acid has unique biochemical and structural properties, different amino acids present in the same position within an amino acid sequence have the potential to change the structure or function of that protein in the cell.

In the hypothetical example shown in figure 5.7, a SNP at position 2 in the third codon of a particular exon has the alleles C and T. The presence of the C allele produces the codon CCG, which codes for the amino acid proline (PRO; mRNA codon CCG) according to the genetic code. Alternatively, the presence of the T allele produces the codon CTG, which codes for the amino acid leucine (LEU; mRNA codon CUG). Individuals who are homozygous for the T allele (in other words, who have the T/T genotype) will have a protein built with the leucine amino acid at this position, while individuals with the A/A genotype will have a protein built with the proline amino acid at that position (C/T heterozygotes would be assumed to have both protein versions produced). Assuming that the amino acid difference between the proteins changes the function of the protein in the cell, the phenotype that is regulated by that protein could be affected by that difference. In the case of figure 5.7, the T allele results in enhanced activity of the protein, causing enhanced bone growth and greater body height.

In order for genetic variation to influence a physical trait, the presence of one allele must affect a gene or protein differently than an opposing allele. There are several ways this can happen, as shown in table 5.1. The example discussed earlier, in which the two alleles at a particular SNP result in different amino acids within the protein sequence, is known as a **missense polymorphism.** Whether such a slight change in the amino acid sequence (i.e., a single amino acid change) will alter protein function is difficult to predict and generally requires direct confirmation through molecular or biochemical tests on the different proteins.

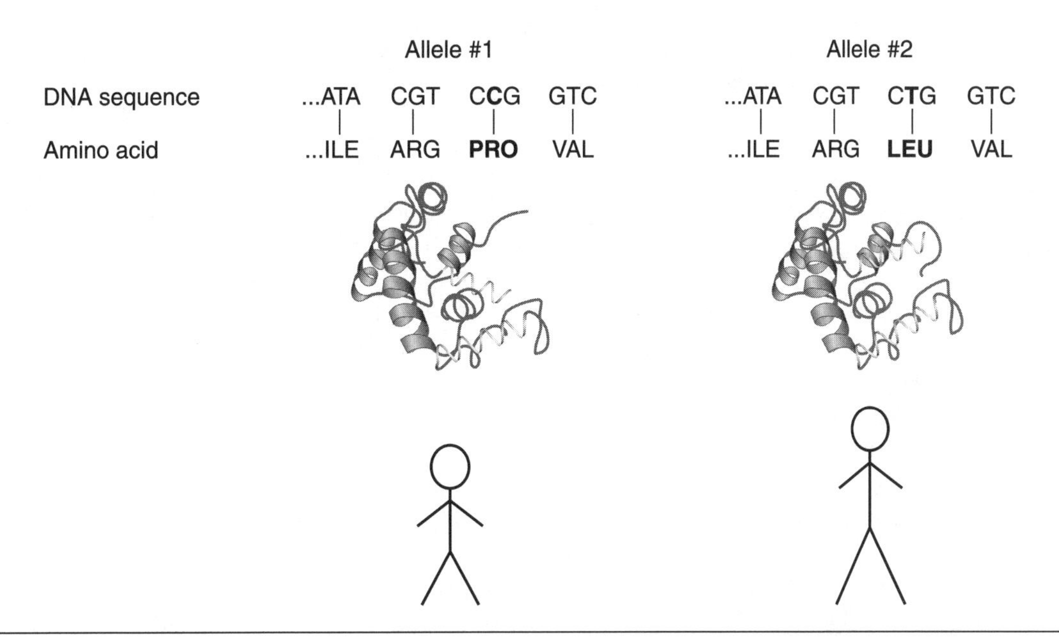

Figure 5.7 An example of how genetic variation can influence a phenotype. In the hypothetical example, a SNP in the coding region of a gene alters the amino acid sequence (i.e., missense polymorphism), resulting in an altered protein structure and function. Because this protein is important for longitudinal bone growth, body height is affected differently depending on the presence of the C or T allele.

Table 5.1 Priorities for Single-Nucleotide-Polymorphism Selection

Type of variant	Location	Functional effect	Frequency in genome	Predicted relative risk of phenotype
Nonsense	Coding sequence	Premature termination of amino acid sequence	Very low	Very high
Missense or nonsynonymous (nonconservative)	Coding sequence	Changes an amino acid in protein to one with different properties	Low	Moderate to very high, depending on location
Missense or nonsynonymous (conservative)	Coding sequence	Changes an amino acid in protein to one with similar properties	Low	Low to very high, depending on location
Insertions/deletions (frameshift)	Coding sequence	Changes the frame of the protein-coding region, usually with very negative consequences for the protein	Low	Very high, depending on location
Insertions/deletions (in frame)	Coding or noncoding	Changes amino acid sequence	Low	Low to very high
Sense or synonymous	Coding sequence	Does not change the amino acid in the protein, but can alter splicing	Medium	Low to high
Promoter or regulatory region	Promoter, 5' UTR, 3' UTR	Does not change the amino acid but can affect the level, location, or timing of gene expression	Low to medium	Low to high
Splice site (intron–exon boundary)	Within 10 base pairs of the exon	Might change the splicing pattern or efficiency of introns	Low	Low to high
Intronic	Deep within introns	No known function, but might affect expression or mRNA stability	Medium	Very low
Intergenic	Noncoding regions between genes	No known function, but might affect expression through enhancer or other mechanisms	High	Very low

UTR = untranslated region.

Reproduced with permission from *Nature Reviews Genetics,* Vol. 3, No. 5, pp. 391-397, copyright 2002 MacMillan Magazines Ltd.

Because each of the 20 amino acids has specific properties, researchers may be able to speculate on whether a certain amino acid change will be more or less likely to affect protein function, depending on the similarity of the two possible amino acids. In other words, when two biochemically similar amino acids are coded for by a missense polymorphism, the protein may not be affected by the change. Conversely, when the two amino acids coded for by the different alleles differ remarkably in their biochemical properties, there is a greater chance for a change in protein function. These two scenarios are sometimes referred to as *conservative* and *nonconservative* missense polymorphisms, with "conservative" referring to two similar amino acids and "nonconservative" referring to two amino acids with different biochemical properties. For example, a SNP resulting in either isoleucine or leucine in the amino acid sequence would be considered a conservative missense SNP because isoleucine and leucine are both considered nonpolar, aliphatic amino acids. Conversely, a missense SNP resulting in proline and lysine would be considered nonconservative, as proline has a cyclic structure and is nonpolar, while lysine is linear and polar. The biochemical properties of each amino acid are described in detail in most biochemistry textbooks.

Table 5.1 shows the many types of DNA sequence variation, their location in the gene or genome, likely effect on the gene or protein, frequency across the genome, and likelihood of significant alterations on the related phenotype. Notice that the likelihood of a strong functional effect (e.g., nonsense or missense polymorphisms) is seen in those variants that occur less frequently across the genome, while the more common variants are those less likely to influence gene or protein function (i.e., low relative risk).

Recall that there is considerable redundancy in the genetic code, so simply changing one allele within a codon does not guarantee a change in the resulting amino acid. When two alleles present within a coding region result in the same amino acid (e.g., AAA and AAG both code for the same amino acid, lysine), the polymorphism is known as a **silent polymorphism**—silent in that there is no change in protein sequence despite the different alleles present at the variant site.

More drastic is the case in which one of the variant alleles in a coding region polymorphism results in one of the three stop codons, which is known as a **nonsense polymorphism.** For example, a SNP with the alleles C and G at the third position of a codon beginning with TA results in either TAC coding for the amino acid tyrosine (mRNA codon UAC) or TAG, which is the stop codon UAG in the mRNA sequence.

KEY POINT

Polymorphisms or mutations can become "functional" when the amino acid sequence is altered by the different alleles found for that sequence variation. Alterations in amino acid sequence can change the structure and function of the resulting protein, which can affect the phenotype associated with that protein.

The presence of a stop codon will necessarily cease protein production at a novel site, in many cases drastically altering the structure of the final protein product. Imagine a nonsense SNP early in the coding region of a gene (sometimes called a **premature stop codon**). The presence of a stop codon at such an early stage of the gene could render the resulting protein completely useless to the cell. While such nonsense polymorphisms are most often seen in the cases of genetic diseases such as muscular dystrophy and cystic fibrosis, examples of nonsense SNPs have also been found with more limited influences on phenotypes. A prime example is a nonsense SNP known as R577X in the skeletal muscle alpha-actinin 3 (ACTN3) gene. This is a true polymorphism, with the rare allele frequency >5% in nearly every population studied (and in some populations >25%); and the alleles at the polymorphism result in either an arginine (R) or a stop codon (denoted as an X). The novel, premature stop codon (X allele) prevents protein translation for that gene copy. Each person carries two copies of the ACTN3 gene, with three possible genotypes: R/R, R/X, and X/X. In X/X homozygotes, where both copies of the gene contain the premature stop codon, *no* alpha-actinin 3 protein is produced. Despite the complete lack of this protein, there is no association with skeletal muscle disease or other overt maladies; however, differences in muscle function and muscle performance in otherwise healthy individuals have been identified depending on the presence or absence of the protein, as will be described more completely in chapter 11.

The gene variants that have been discussed so far are primarily those that have direct influences on proteins (i.e., are located in the coding regions of the gene), either by changing an amino acid or by inserting a novel stop codon early in the gene sequence. While such genetic variants are potentially quite important for altering protein function, the amount of DNA dedicated to exons within the human genome is estimated at less than 3%. So, although these variants can be important, we are far more likely to find variants in the noncoding regions of the genome, including the regions up- and downstream of a gene and within the introns between exons. For polymorphisms in these regions, the potential influence on gene and protein function lies not within the amino acid structure itself, but rather with the transcription of the gene and the modification of the resulting mRNA.

For example, SNPs that fall within the promoter region of a gene may alter the sequence of a transcription factor binding signal, which may prevent the transcription factor from binding to the promoter when one of the two alleles in the SNP is present. In that case, the presence of one of the alleles may result in lower or higher gene transcription rates compared to the other allele, potentially altering how much mRNA and protein are available within the cell. Similarly, SNPs that fall within the intronic regions or the downstream regions of genes can influence the modification of the mRNA sequence, such that the degradation of mRNA may occur faster or more slowly depending on the allele present at the particular site. There are numerous examples of how various polymorphisms in different gene regions can potentially influence gene function with effects on the related phenotype. For our purposes, we will rely on the general traits of polymorphisms shown in table 5.1. In the table, the likelihood of effects on gene or protein function is shown for different types of polymorphisms in several different genomic locations. This is a useful guide for determining the potential importance of a particular polymorphism compared to another within the same gene, which is relevant when one is selecting different polymorphisms to study within a gene.

HOW DO POLYMORPHISMS ARISE? GENETIC VARIATION AND EVOLUTION

All through this chapter, we've discussed the idea of genetic variation among humans—but why does such variation exist, how do new alleles develop, and why are they present in so many individuals? The answer to all of these questions is built on a foundation of understanding the basics of **evolution,** which is simply the idea that a DNA sequence can and does change over time (defined as generations of a species). In this context, evolution does not mean that species are changing to become different species (though that can happen), but rather that various environmental factors can alter DNA sequence in one individual of a species, and those sequence changes can become permanent and widespread in that individual's population over many, many generations. It is this basic premise upon which we can discuss the specifics of polymorphisms in the human species.

In addition to simple mistakes by the cell machinery used to replicate DNA prior to cell division (the DNA polymerase enzyme is ~99.99% accurate), various environmental factors, known as **mutagens,** can result in the mutation of DNA sequence. These mutagens include various chemicals, some viruses, and radiation, including X-rays. Any of us can be exposed to such factors, though there is a strong element of chance regarding whether or not a DNA mutation will result from such an exposure. Moreover, if DNA is mutated in response to the exposure, that DNA mutation will occur only in the cell that is exposed to the mutagen. If one of your liver cells gains a mutation in a single nucleotide of the DNA sequence, this may or may not have any consequences for you in your lifetime. In other words, the liver cell may die as a consequence, with no effect on overall liver function; it may continue to function normally through its normal cell life cycle; or it may become cancerous, growing in an unpredictable and abnormal way, which could have drastic consequences for both the liver and whole-body homeostasis. But none of this leads to a mutation with *evolutionary consequences,* because the mutation did not occur in a sex cell. Only the sex cells are capable of transferring DNA sequence to the next generation, so it is these cells that must carry a new mutation in order to potentially affect the DNA sequence of the next generation (and then possibly an entire population over many generations).

If a sex cell is mutated such that a new allele is present in the given individual but not in anyone else in the population, that new DNA sequence mutation can be passed along to that individual's offspring. Once the cell becomes part of the fertilization and development process in the offspring, all of the body cells of the new offspring will carry the new mutation, as the sex cells form the basis for all other cells in the new organism. Now the question becomes, What is the consequence of this new mutation (i.e., allele)? The possible consequences follow exactly our discussion of the consequences of genetic variation in general, with intronic polymorphisms having low probabilities of altering gene or protein function, and coding region polymorphisms having considerably higher probabilities of altering protein function. Thus, if the new allele is present in a coding region, resulting in a new amino acid in the coded protein sequence (or more severely, a new stop codon), the consequences could be significant for that individual, with the new protein potentially altering one or more phenotypes in comparison to other population members.

There are many consequences that a new mutation could have in an individual: (1) no effect, positive or negative, on the individual, with either no change in any phenotype or with such minimal changes in the phenotype that the life of the individual is not affected; (2) a negative effect, such that the new allele results in a defect that prevents normal function, perhaps as severe as death before birth or as minimal as mild cognitive or physical impairment or disease susceptibility; or (3) a positive effect, such that the new allele results in a positive impact on a phenotype, for example providing slight improvements in cognitive function, muscle strength, carbohydrate metabolism, and so on. Depending on the consequences of the new allele, the ability of the individual to survive to adulthood and reproduce, a concept known as **evolutionary fitness,** may be affected. Notice that the use of the term "fitness" is different here from that in the typical exercise science context; evolutionary fitness has to do with an individual's ability to survive to reproduce and pass on his or her DNA sequence, which is different from aerobic or muscular fitness.

If the consequence of the new allele is *negative,* it is possible that the individual will not survive to adulthood or, if the individual reaches adulthood, will not reproduce because of difficulties in attracting a mate. On the other hand, if the mutation is *neutral* (having no effect) or *positive* (conferring some advantage to the individual), the individual will have no disadvantage compared to others in his or her group surviving to adulthood and reproducing, and may have an advantage compared to others (in the case of a positive effect). Thus, that individual can go on to reproduce, with the possibility of passing along the new mutation to offspring. Especially if the new mutation provides some advantage for the individual, the odds of passing the allele on to new offspring are increased, as presumably the individual will have greater survival chances (i.e., greater evolutionary fitness) and may thus be more attractive in terms of mate selection, with greater opportunities for reproduction. If these new offspring carry the advantageous allele, then they too will have the same phenotypic advantage as their parent, providing the same survival and reproduction advantages. Over many generations, what started as a new mutation in a single individual can spread to being carried in many individuals in the population (or perhaps all individuals, if the effect is especially advantageous and enough time has passed). These are the basic ideas behind Darwin's concept of **natural selection,** which he proposed as a mechanism for evolution. In other words, natural forces such as environmental change (affecting survival of certain individuals) and mate selection patterns (affecting reproduction rates by certain individuals) can alter DNA sequence in a population over many generations. More information on evolutionary forces can be found in appendix B, "Evolution and Hardy-Weinberg Equilibrium."

> **KEY POINT**
>
> Evolution is simply a change in DNA sequence in a population over successive generations. Those DNA sequence changes can be neutral (no effect on a phenotype) or may alter a phenotype with either positive or negative consequences for individuals in the population.

SUMMARY

Genetic influences are first indicated by the presence of familial aggregation or similarity for a phenotype. When families, which share considerable fractions of their DNA sequence, are similar for a trait of interest, this indicates that either genetic factors, shared environmental factors, or both are important for that trait. To estimate the fraction of this variance attributable to genetic factors (known as the heritability), specific family studies are performed and correlations among different family members are compared. When heritability has been shown for a particular trait, this indicates that genetic variability somewhere in the genome is influencing that trait. Identifying which genes are potentially important is the topic of chapter 8; but in general, genetic variation simply means that nucleotide differences among the DNA sequence of individuals can influence the function of genes and their proteins. Many types of genetic variation exist, including single-nucleotide polymorphisms (SNPs), insertion/deletion polymorphisms, and repeat polymorphisms. Within these polymorphisms, different alleles are found in different individuals, and those alleles can influence the gene and its protein by altering amino acid sequences or influencing gene regulation. These issues are examined in greater detail in the next chapter. Polymorphisms arise in a population through various evolutionary forces, some of which may result in a novel DNA sequence mutation that is passed along to many individuals over many generations.

KEY TERMS

allele
allele frequency
correlation
D allele
dizygotic twins
evolution
evolutionary fitness
familial aggregation
frameshift mutation
fraternal twins
genetic variation
genotype
heritability
heterozygote
homozygote
I allele
identical twins
insertion/deletion polymorphism
microsatellite repeat
missense polymorphism
monozygotic twins
mutagen
mutation
natural selection
nonsense polymorphism
polymorphism
premature stop codon
repeat polymorphism
silent polymorphism
single-nucleotide polymorphism (SNP)

REVIEW QUESTIONS

1. How can twin pairs be used to determine heritability of a trait?
2. What is genetic variation, and what types of variation are typically seen in the genome?
3. How does a mutation differ from a polymorphism?
4. In general, how can DNA sequence variation influence the function of a gene or protein?
5. Using the genetic code in table 3.1, identify possible nonsense, missense, and silent single-nucleotide polymorphisms. How could you label those polymorphisms?
6. Describe DNA sequence mutation and the process of DNA sequence evolution. In other words, how can a polymorphism arise?

CHAPTER

6

GENETIC VARIATION AND DISEASE

As outlined in the previous chapter, genetic variation can result in differences in gene and protein function among individuals, causing differences in physical traits. While these differences are often minor, resulting for example in the slight differences we see in the physical appearance of people, in some cases genetic variation can have dramatic consequences for a phenotype, resulting in overt disease or a strong disease predisposition for an individual. In this chapter, we explore the two major categories of genetic disease: Mendelian disease and complex disease. And we'll see that understanding the role of genetics in disease allows an understanding of how genetics can interact with nondisease phenotypes as well, for example influencing physical fitness and sport performance. In chapter 12, we will discuss the specific use of genetic information in medicine, health care, and exercise prescription.

MENDELIAN DISEASE GENETICS

We start our discussion of disease genetics by focusing on those phenotypes influenced by single genes, which are often first observed in families with unique disease predispositions. As we'll see, what may seem an obscure association between a gene mutation and a rare disease or other phenotype can actually provide important insights into both common diseases and general physiology.

Family-Based Disease Patterns

What is one of the first forms you complete after arriving at a doctor's office? After your name and other basic information, one of the first items of interest for health care providers is your family medical history. Starting with their mother and father, and extending to siblings, grandparents, and even more extended family, patients are asked to report on various disease patterns that may exist within their family tree. The basis for this very common practice in health care is the shared genetic and environmental backgrounds of families. When a grandparent carries a genetic variant known to influence disease, that same genetic variant can be passed along to multiple generations, thus passing along the inherent disease risk as well. Knowing a patient's family medical history can help point to such genetic or environmental associations (or both), helping health care providers to devise more individualized prevention and treatment plans.

The most dramatic example of such family-based disease patterns is the case in which a single base mutation within the DNA sequence results in a disease phenotype in all carriers of that mutation. These **single-gene diseases** or so-called **Mendelian diseases or traits** are often quite rare but can have dramatic influences on a particular trait. For example, single genes and mutations within those genes have been identified for a number of rare diseases, including familial hypercholesterolemia, Marfan syndrome, cystic fibrosis, and Tangier disease. A more detailed example relating to muscular dystrophy is presented in the Special Focus section in this chapter. In Mendelian genetic disease, only one gene is important for the disease phenotype, often with little or no environmental influence. Hundreds of such single-gene disorders exist, though the exact genetic mutation causing the disease is not known in all cases. As discussed previously, the likelihood of inheriting such single-base mutations is predictable based on Mendelian inheritance patterns. So, when one parent carries a dominant disease-causing mutation with a 50% chance of inheritance for each child, we can predict the risk for disease development in each child (similarly for a child of parents each carrying recessive disease gene alleles). More importantly, for diseases for which the underlying gene mutations are known, those mutations can be specifically screened within children to more accurately determine disease risk, allowing for the possibility of preventive care prior to the onset of disease symptoms.

KEY POINT

Mendelian diseases are those influenced by a single gene, such that presence or absence of a disease gene will result in presence or absence of the related disease.

Rare genetic mutations such as those just mentioned are not necessarily tied to negative outcomes. In fact, as noted in chapter 5, most gene variants occur in noncoding regions of the genome, with less of a chance for negative influences on a protein. And even rare mutations within coding regions cannot be assumed to have a negative outcome. An excellent example of this comes from the sporting world. Throughout the 1960s, a Finnish cross-country skier excelled in his sport at the international level, winning several world championships and Olympic medals. While always dogged by accusations of blood doping for performance enhancement, Eero Mäntyranta always claimed innocence, despite the fact that his training methods were no different from those of his slower peers. Years later it was determined that Mäntyranta did have an advantage over his peers, though it was not something he had control over. Mäntyranta and his family were identified as having abnormally high levels of red blood cells and hemoglobin, consistent with higher red blood cell production. Recall that red blood cells, or RBCs, are critical for the transport of oxygen from the lungs to the working tissues; their production is stimulated in the body by the hormone erythropoietin (EPO).

Mäntyranta and his extended family became the focus of a research study aimed at identifying the basis for these clinical abnormalities. Ultimately, a rare genetic mutation was identified within those members of the family carrying the abnormally high RBC and hemoglobin levels. That mutation, a rare A allele (vs. the typical G allele) within exon 8 of the erythropoietin receptor (EPOR) gene, results in a stop codon leading to a shortened receptor protein. Remarkably, this shortened receptor is more active in response to the binding of EPO, thus causing enhanced RBC production and hemoglobin-carrying capacity. Fortunately for the Mäntyranta family, this mutation does not cause negative disease outcomes, but rather provides an advantage in the arena of aerobic exercise performance. So, while Finland's famous skier appears not to have illegally enhanced his performance, he did have an uncommon genetic advantage compared to his peers.

Studying the Mechanisms

Though single-gene disorders with drastic disease phenotypes are not the focus of interest for this particular text, the mechanisms for single-gene disorders are instructive for more common disease phenotypes such as type 2 diabetes, obesity, and hypertension, as well as for our understanding of physiology in general. For example, the specific mutations that underlie Mendelian diseases are often instructive for the biological pathways in which those proteins normally function. In other words, when phenotype or disease traits are associated with a specific missense or nonsense mutation in a protein, a window is opened for researchers struggling to understand the normal physiological function of that protein or signaling pathway.

Another instructive area regarding Mendelian diseases centers on the fact that the predictions of who will acquire a genetic disease are not always perfect. In other words, while a person carries a mutation known to cause disease in the vast majority of carriers, that person may be spared from disease. This concept is known as **penetrance,** or the proportion of individuals with a given genetic mutation that exhibit the disease phenotype. The penetrance of a disease gene mutation is not always 100%, as would be predicted; rather, some individuals are miraculously spared from exhibiting disease symptoms.

How an individual could carry such a mutation without developing disease will depend on the specific gene, specific mutant allele, and the related disease phenotypes; but for our purposes the general idea is that *other factors* in addition to the primary disease gene can be influential in determining the presence and progression of disease. Those other factors could be other genes and genetic variants of minor influence, such that unique combinations of other genetic variants in other genes may somehow interact with the primary gene mutation, thus preventing its negative influence. Alternatively, lack of disease in a gene mutation carrier might indicate the interactive effect of environmental factors on the gene mutation, such that specific environmental factors may be necessary for the development of disease, or specific environmental factors may prevent disease in otherwise susceptible people.

An example of this is phenylketonuria (PKU), which is the term for drastically elevated plasma levels of the amino acid phenylalanine owing most commonly to a gene mutation in the phenylalanine hydroxylase enzyme (PAH) that prevents normal enzyme function. In other words, without the enzyme needed to convert phenylalanine to tyrosine in the body (a consequence of the gene mutation), phenylalanine levels increase in the blood, with the consequence of mental retardation. Despite the severe risk associated with this genetic mutation, the disease phenotype cannot develop without the ingestion of phenylalanine in the diet. Thus, PKU is not a simple Mendelian disorder, but rather an example of a *gene × environment interaction.* That is, the gene mutation by itself does not lead to disease; rather, the intake of phenylalanine in the diet is required to interact with the mutant allele in the PAH gene. Thus, the most common, successful treatment of PKU is a low-phenylalanine diet, minimizing plasma levels of phenylalanine and generally preventing the cognitive abnormalities associated with untreated PKU. Phenylalanine is now commonly screened for in infants using a genetic test performed soon after birth, allowing an early modification of the infant's diet in the case of a PAH gene mutation.

The idea that multiple genes or environmental factors (or both) may influence the presence or progression of a genetic disease brings us to the primary focus of interest for this text: complex phenotypes.

SPECIAL FOCUS
Muscular Dystrophy

A classic example of Mendelian genetic disease is muscular dystrophy, an inherited genetic defect that results in a diverse group of muscle disorders. All of these disorders are characterized by progressive muscle weakness and atrophy caused by various genetic mutations. In this section we will focus on mutations within the dystrophin gene, which codes for the skeletal muscle protein dystrophin (other genes are involved in some types of muscular dystrophy). Mutations in the dystrophin gene are linked with Duchenne muscular dystrophy (DMD), the most severe form of muscular dystrophy. Duchenne muscular dystrophy is most commonly found in boys, with symptoms typically beginning before the age of 5 years and resulting in death before the age of 20 years.

The dystrophin gene is the largest gene in the human genome, consisting of 2.3 million nucleotides with 79 exons. The gene is located on the X chromosome, so men have only one copy of the gene compared to two copies in women. Mutations in the dystrophin gene, especially in the form of deleted exons, can be nonsense mutations, resulting in the lack of a functional dystrophin protein. The dystrophin protein is an important cytoskeleton protein in skeletal muscle cells, specifically located beneath the cell membrane where it binds with other proteins to form the dystrophin-glycoprotein complex. This complex of proteins is important for maintaining the structural integrity of the muscle cell membrane, especially during contraction, and it may have other roles in skeletal muscle cells as well.

The inheritance of DMD follows many of the classical rules of Mendelian genetics. Because females carry two copies of the gene, a single mutated version rarely results in disease for a female since the functional gene copy can provide the information needed to produce a functional dystrophin protein. Males, who have only one copy of the gene, are destined to show disease symptoms when carrying a

(continued)

SPECIAL FOCUS

(continued)

rare mutation. Since male carriers of a mutated gene will rarely live to reproductive age because of the disease symptoms, they are not likely to pass along such a mutation to their offspring. (Recall that only female offspring would be at risk for receiving an X chromosome gene mutation from their father.) So disease screening for DMD is focused on mothers, as only mothers can pass along a mutated gene copy of the dystrophin gene to their offspring, but they are typically disease free and thus require screening to detect a dystrophin gene mutation. Female offspring can be carriers, like the mother and also without symptoms, but male offspring, who carry only one X chromosome, will have disease symptoms. Thus, muscular dystrophy is known as an *X-linked genetic disease.*

Remarkably, though, simple Mendelian inheritance of dystrophin gene mutations does not explain all cases of the disease. In fact, only about two-thirds of cases are attributed to a mother passing on a mutated gene copy. The other one-third of cases are attributed to new mutations in the dystrophin gene that occur in the female sex cells. The large size of the dystrophin gene makes it susceptible to new genetic mutations; the odds for a mistake in DNA replication over 2.3 million nucleotides are proportionally higher than for genes of 23,000 nucleotides (roughly the average length of a human gene). Thus, the dystrophin gene is no more susceptible to mutation than any other gene, but there is simply more of the gene to be copied, making errors more likely.

COMPLEX DISEASE

While any disease can be thought of as complex, for our purposes **complex disease** is defined as a trait in which multiple genes and environmental factors interact to contribute to the presence or absence of the disease phenotype. Note that genetic factors and environmental factors are emphasized equally here, as both play a critical role in complex disease progression. Complex diseases are also called *multifactorial* or *combinatorial,* owing to this combination of multiple interacting factors. When we are referring to the contribution of genetic factors, complex traits are also known as **polygenic** ("poly" meaning many, "genic" referring to genes).

Note also that while the term complex *disease* has been used to introduce this section, complex *phenotype* could easily have been the term of focus. Ultimately, complex diseases are complex phenotypes that fall beyond some threshold defined as the limit of a "normal" or healthy phenotype. When a phenotype is above or below this critical threshold, the phenotype is labeled as a disease rather than as part of the normal distribution of phenotype values. While body fat percentage can be thought of as a complex phenotype, once that percentage reaches a certain high threshold (say 25% in men and 32% in women), the phenotype is referred to as the disease phenotype obesity, which is discussed in depth in the Special Focus section. Similar examples could be provided for any number of complex phenotypes where abnormal levels are considered disease, including blood pressure (hypertension), glucose levels (diabetes), or cognitive function (dementia). Complex phenotypes are often the focus of study within the contexts of medicine and exercise science, with the emphasis on prevention or treatment of the disease end of the phenotype continuum. Thus, while $\dot{V}O_2max$ is considered a complex phenotype that could be studied within the context of sport performance at the high end of its continuum, it can also be targeted as a major risk factor for cardiovascular disease and mortality at low values.

Similar to Mendelian disease, complex disease can be studied in families, and estimates of heritability can be obtained. Unlike the situation with Mendelian diseases, however, many genes and many environmental factors are influencing the complex trait. The number of genes and environmental factors, as well as the extent to which any one factor contributes to a trait, will depend on the specific phenotype of interest. As was shown in the pie chart example in figure 5.3 (chapter 5), if we consider that all of the variability of a complex trait can be divided roughly into environmental versus genetic influences (estimated by heritability studies on that trait), within each of those major sections we see that the number of factors and the influence of each of those factors can be variable. For instance, think of the major environmental factors important to blood pressure: smoking, physical activity, diet, and so on. All of these factors would fit within the environmental side of the blood pressure (or hypertension) pie chart. Similarly, blood pressure has been shown to be heritable and is considered a complex genetic trait, in that mutations in many different genes have been shown to result in Mendelian-type hypertension in various families—suggesting that multiple genes in multiple physiological pathways can influence blood pressure across different individuals. This same scenario can be used to describe the many other common complex phenotypes of interest in medicine and exercise science, with multiple genetic factors and multiple environmental factors interacting to determine the presence and severity of disease.

Imagine now the distribution of genetic variants within the genetic factors side of our phenotype pie chart. For a complex trait, the likelihood is that many genes, potentially tens to hundreds, could be contributing to a particular phenotype of interest, and the importance of each gene could be different. In other words, in a complex trait where 25 genes and gene variants are influencing a phenotype, all but four of those genes could be contributing less than 2% of the variability of the trait, while each of the four remaining genes has a greater contribution. The number of genes and their individual importance would of course be different for each complex trait of interest, owing to differences in the physiology underlying each trait.

KEY POINT

Complex phenotypes are influenced by multiple environmental factors and multiple genetic factors, making the identification of important genetic factors challenging.

An important caveat to this discussion is the idea of what are often called **intermediate phenotypes** or **subphenotypes** (and occasionally referred to as endophenotypes). An intermediate phenotype is one that underlies a broader phenotype of interest but is influenced by fewer genes or environmental factors. For example, blood pressure is a common phenotype of interest given its importance to cardiovascular health, but blood pressure is an extremely complex and broad phenotype to study. Blood pressure is defined by multiple physiological systems in the body, and several environmental factors can influence all of those systems, making the regulation of blood pressure extremely complex. If we were to imagine a pie chart of the various genes and environmental factors for the phenotype of blood pressure, we could imagine many genes of relatively small effect on the genetic factors side of the pie, and similarly several environmental factors contributing to the environmental factors side of the pie.

A possible alternative to studying such a broad phenotype would be to select an intermediary phenotype underlying blood pressure that will be influenced by fewer factors. Some examples of intermediate phenotypes for blood pressure could be blood levels of the components of the renin-angiotensin system, blood flow and vasodilation capacity, and renal sodium excretion or salt sensitivity. While all of these subphenotypes are important for the regulation of blood pressure, the number of genetic and environmental factors important for each of these more targeted phenotypes is likely to be considerably smaller than that for blood pressure as a whole. The number of genetic and environmental factors for researchers to identify will be smaller and the identification potentially more straightforward. An example is shown in figure 6.1. Each oval shape represents an intermediate phenotype underlying a broader phenotype (e.g., blood pressure). While the broad phenotype (circle) in this example is influenced by four genes (triangles) and four environmental factors (rectangles), the intermediate phenotypes underlying the broad phenotype are each influenced by only one to two genes and environmental factors, making the identification of genetic and environmental factors potentially easier for the intermediate phenotypes than for the broad phenotype of interest. The problem presented by intermediate phenotypes is that they are often more challenging and expensive to measure in humans, so researchers must find a balance between the limitations of intermediate phenotype measurement (e.g., cost, discomfort to subjects) and the limitations of phenotypes so broad that many, many factors will exert only a minor influence.

The complexity of these multifactorial, complex phenotypes can be truly overwhelming, and it is important to take a step back and gain some perspective on this issue. As will be discussed in more detail in chapter 12, the future use of genetic information

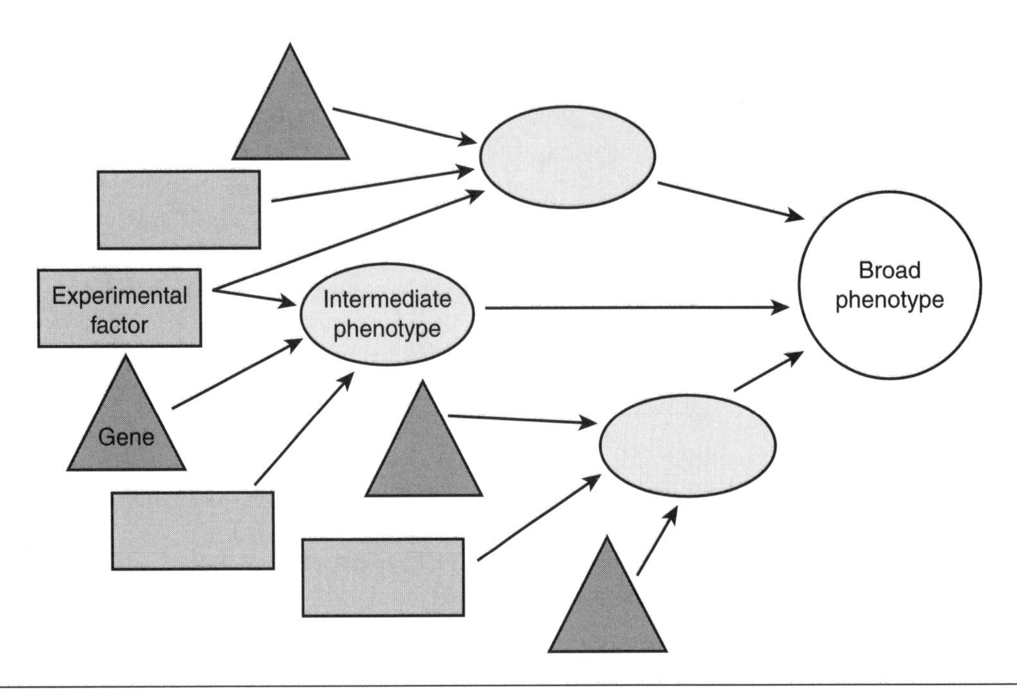

Figure 6.1 Intermediate phenotypes or subphenotypes underlying a single, broad phenotype.

in medicine and exercise science will require discrete, tangible information about the importance of a particular gene variant for a particular phenotype. In the case of the pie chart example in figure 5.3 (chapter 5), while many genes are shown as important to the phenotype, only a small fraction are shown to have a strong influence individually. The focus of researchers in the area of genetics is in identifying those major genetic influences, as these will be the ones of likely value in the clinic. So, while the complete listing of genes, environmental factors, and their interaction may seem overwhelmingly complex, the goal is to identify the major influences underlying any particular phenotype for targeting by the medical or exercise science community.

GENE × ENVIRONMENT INTERACTIONS

Complex phenotypes are made more complex by the fact that genetic factors and environmental factors can interact (i.e., gene × environment interactions) such that any one gene or environmental factor by itself may have minimal influence, but acting together the various factors can have a large influence on a phenotype. We've already noted one such example, that of PKU, which is prevalent only in individuals with a genetic defect *and* high dietary intake of the amino acid phenylalanine.

To solidify the concept of gene × environment interactions, let's discuss a few hypothetical examples before examining some findings from the research literature. Consider a sedentary lifestyle (i.e., physical inactivity) and cardiovascular (CV) disease. While physical activity is a potent means of preventing CV disease in many individuals, CV disease-related deaths do occur in individuals with high levels of physical activity at the same time that others live completely sedentary lives but remain free of CV disease symptoms. One explanation for such discrepancies would be unique gene × environment interactions, such that the environmental factor (physical inactivity) is especially harmful in individuals with specific genetic variants while in other individuals the harm posed by the environmental factor is offset or negated by other specific genetic variants. In both cases, genetic and environmental factors are interacting to result in a measurable influence on the phenotype.

A similar example can be imagined for lung cancer. We know that not everyone who smokes develops lung cancer, though smoking is the greatest risk factor known for the development of lung cancer. Moreover, anecdotal evidence suggests that some individuals can live long lives without developing lung or other cancers despite a lifelong smoking habit, while others develop lung (or mouth or throat) cancer after only a short exposure to tobacco use. In the same way, a high-salt diet can result in hypertension in some individuals

SPECIAL FOCUS
Obesity as a Complex Trait

A classic example of a complex trait is body fat, high levels of which are defined as obesity. Obesity is commonly defined as a body mass index (BMI; kg/m^2) value greater than 30 kg/m^2, and approximately one-third of U.S. adults were estimated to be obese at the turn of the 21st century. Moreover, nearly two-thirds of U.S. adults were considered overweight (BMI $\geq$ 25 kg/m^2) at that time. Why is this a concern? Obesity is linked to the development of cardiovascular disease, type 2 diabetes, stroke, cancer, and other disorders, with associated health care costs running in the billions of dollars annually. With the trends pointing to even higher rates of obesity in the near future, there is considerable effort on the part of many organizations to slow weight gain and reduce obesity rates.

Complex traits encompass both genetic and environmental factors as important contributors to development of the trait, and obesity certainly qualifies as such a complex trait. Considerable evidence shows that caloric intake and physical activity levels are two very important environmental factors that contribute to changes in body fat levels, and other environmental factors have been identified as influencing diet and exercise habits. With regard to genetic factors, consider our typical strategies for identifying a genetic contribution to a trait: familial aggregation and heritability studies. For obesity, many studies have shown familial aggregation for various body composition measures, and those studies have been followed up by heritability studies showing that genetic factors contribute an estimated 40% to 80% of the total phenotype variance for body fat (different studies report different estimates). In addition, many single-gene mutations have been found to result in obesity (i.e., Mendelian-type mutations in single families or individuals), such as mutations in the leptin and the leptin receptor genes. The sum total of all of this evidence points to the contribution of both genetic and environmental factors to the development of obesity, thus making it a complex trait. In fact, in the latest version of the Human Obesity Gene Map (Rankinen et al., 2006), 127 genes have been positively associated with obesity-related phenotypes. While 22 of those genes have been positively associated with obesity-related phenotypes in at least five different studies, no genes had been conclusively identified as key contributing genes for obesity.

The bottom line is that both genetic and environmental factors (and their interactions) are likely to explain obesity in any one person. But the very large increases in rates of obesity over the past few decades in the United States cannot be explained solely by genetic factors. The human genome sequence has not changed enough over the past 100 years to explain the incredible increases in obesity rates. Rather, environmental factors have changed; especially there has been a drastic reduction in physical activity levels since the early 20th century. But the environmental factors of diet and physical activity may be more important in individuals with certain genetic profiles, such that specific gene $\times$ environment interactions are critical to the development of obesity. One argument is that the human genome is not adapted to a sedentary environment, having evolved over centuries of hunter-gatherer living, so that low levels of physical activity are harmful in two ways: (1) the simple reduction in energy expenditure leading to a positive energy balance and (2) the loss of key gene $\times$ physical activity interactions important to the maintenance of body weight and other health-related phenotypes.

Obesity researchers are seeking the answers to these questions with the ultimate goal of developing treatment and prevention strategies that will improve health and decrease health care costs. Until that goal has been realized, however, the strategies of increasing physical activity and reducing dietary intake (i.e., altering environmental factors) independent of genetic information will remain the primary tools for weight management. As a respected geneticist, J.B.S. Haldane, is said to have stated in the early 20th century, "We do not know, in most cases, how far social failure and success are due to heredity, and how far to environment. But environment is the easier of the two to improve."

without the same effect in others, while reducing salt in the diet may have either a positive benefit or no effect on blood pressure depending on the genetic profile of a particular person. All of these hypothetical examples involve an environmental factor that interacts with a specific gene variant, with the combination influencing a phenotype or disease outcome.

KEY POINT

Gene × environment interactions occur when a specific environmental factor influences a phenotype differently depending on the different alleles present within a gene important for that phenotype.

Such gene × environment interactions are challenging to identify for researchers, but as the tools of genetics research improve, more and more will be identified in the future. Some interesting examples have already been reported, and we'll examine a few of them briefly here.

In the first example, an association between genetic variation in the apolipoprotein E (ApoE) gene and blood cholesterol levels has been observed in several research studies, such that carriers of the ApoE $\epsilon4$ allele tend to show higher levels of low-density lipoprotein (LDL) cholesterol (i.e., "bad" cholesterol) compared to either $\epsilon2$- or $\epsilon3$-allele carriers. This relationship, however, appears to be dependent on the intake of dietary fat. If dietary fat is lowered, $\epsilon4$ carriers show reductions in cholesterol levels to a greater extent than seen in $\epsilon2$ or $\epsilon3$ carriers. Other studies have shown that the type of dietary fat may also interact with ApoE genotype to influence blood lipids. In this example of gene × environment interaction, dietary patterns and ApoE genetic variation appear to interact to influence blood cholesterol levels. This example is described in more depth in chapter 12.

A second example involves body fat levels and their response to physical activity. The use of exercise as a means of reducing body weight and body fat is well known. Researchers studying the genetic aspects of obesity observed that a specific genetic variation in an adrenergic receptor gene increased the risk of obesity in men. But, in those men who maintained even modest physical activity levels, the genetic predisposition to obesity was erased, such that physically active men with the risk allele had body composition measurements similar to those of sedentary men without the risk allele. In this case, physical activity is interacting with a specific genetic factor, thus altering disease risk.

Another example of gene × environment interaction involves the interaction of physical activity levels and genetic factors with the risk of developing Alzheimer's disease. In this example, genetic variation in the ApoE gene that significantly increases the risk for the development of Alzheimer's disease appears to be counteracted by physical activity. This example is discussed in detail in chapter 11.

Because of the complexity of all of the phenotypes that were highlighted in these few examples, additional research will be needed to confirm the observed interactions. If confirmed, then the specific gene × environment interactions may someday be exploited in the treatment of the various conditions, which is the idea behind personalized medicine, discussed in chapter 12.

GENE × GENE INTERACTIONS

An idea similar to gene × environment interaction is that of ***gene × gene interaction***, which is also a part of complex, polygenic phenotypes. In this case, alleles in one gene would interact with alleles in other genes in an interaction with unique consequences for a phenotype. For example, let's consider a gene that codes for a transcription factor, a protein important for regulating gene expression. If an allele in the transcription factor gene affects the function of the transcription factor protein (e.g., missense polymorphism), then when that transcription factor protein interacts with the DNA of its target gene, the consequences for gene expression of that second gene will depend on which allele was present in the transcription factor gene. Now, if the second gene contains a polymorphism where the transcription factor binds, such that one allele allows tighter binding of the transcription factor to the DNA than the other allele, then there is a combination of polymorphisms in both genes that ultimately influence the transcription of the second gene. In other words, predicting the transcription level of the second gene requires knowing which alleles are present in both the transcription factor gene and the second, target gene's promoter region; and different combinations of alleles in these two genes will have different consequences on gene expression of the target gene. This is known as a *gene × gene interaction.* While no concrete examples of gene × gene interactions have been shown, these interactions are assumed to play at least a minor role in the regulation of complex phenotypes, and perhaps a major role in some cases.

SUMMARY

Mendelian or single-gene disorders are those caused by a genetic variant in a single gene, such that presence of the mutant allele in that gene results in the presence of the disorder. Environmental factors are not typically important to the development of Mendelian genetic diseases, though when they are, the penetrance of a genetic disease can be influenced. Not all Mendelian traits are negative, as seen with the example of the skier Eero Mäntyranta. Complex phenotypes, on the other hand, are influenced by multiple genetic and environmental factors and their interactions. Obesity, blood pressure, and $\dot{V}O_2max$ are excellent examples of complex phenotypes. Studying such complex phenotypes can be challenging, and one option for researchers is to focus their efforts on intermediate phenotypes—those that underlie the broader phenotype but are influenced by fewer genetic and environmental factors. Genes and environmental factors can also interact with unique consequences for a phenotype, as was shown for PKU and several other examples.

KEY TERMS

complex disease
intermediate phenotype
Mendelian disease or trait
penetrance
polygenic
single-gene disease
subphenotype

REVIEW QUESTIONS

1. How do Mendelian traits differ from complex phenotypes?
2. For a specific genetic disease that results from an autosomal dominant gene mutation, only 9 of 10 carriers of the dominant mutation develop the disease. What is the penetrance of the disease, and why might some disease gene carriers be spared disease symptoms?
3. What is an intermediate phenotype or subphenotype? In what ways are such phenotypes both easier and more challenging to study for researchers interested in identifying genetic and environmental factors important to a broader phenotype?
4. Identify a complex phenotype that you are interested in but that was not discussed in this chapter (e.g., lung function, muscle strength, body height). What are some environmental factors that might be important for your trait of interest? What are potential intermediate phenotypes that could be studied instead of the broad phenotype (assuming that genetic factors are important)?
5. What is a gene × environment interaction, and how can such an interaction influence the presence of disease?

CHAPTER

7

LINKAGE DISEQUILIBRIUM, HAPLOTYPE, AND ENVIRONMENTAL INTERACTION

As discussed in the previous chapter, complex traits are influenced by multiple genes and polymorphisms, in combination with multiple environmental factors. More and more evidence is pointing to the idea of multiple, interacting polymorphisms, even in the same genes, as important to the regulation of complex phenotypes. Here, we discuss in depth the issue of multiple polymorphisms within a gene and how their alleles, working in combination, may affect phenotype variation.

RECOMBINATION REVISITED

To set the stage for understanding the interaction of multiple polymorphisms in a gene or gene region, we must revisit the issue of chromosomal recombination, covered in chapter 4. Recall that when the sex cells undergo meiosis in preparation for fertilization, the DNA of each parent is mixed, such that the DNA of each sex cell contains DNA sequence that is unique from that of the parent's somatic cells. This semi-random process involves the chromosomes aligning and undergoing crossover, such that entire segments of DNA are exchanged between homologous chromosomes, thus forming unique combinations of the parental DNA strands.

The key piece of information needed for expanding our definition of genetic variation is that chromosomal recombination involves the exchange of segments or blocks of DNA sequence, known as **recombination blocks.** The DNA is not exchanged letter by letter across homologous chromosomes; rather, several thousand, perhaps even several million, letters in a single block of DNA sequence are exchanged between the paired chromosomes.

With this idea as background, let's consider figure 7.1. Here, we see that DNA can be thought of as not just a sequence of individual nucleotides strung together along a chromosome, but instead as small blocks of DNA sequence, each block composed of numerous individual letters, with many blocks making up each chromosome. This block structure of DNA is shared among all individuals in a population, such that the block sizes and locations (i.e., block boundaries) are the same, but the actual DNA sequence contained within each block may differ slightly depending on the specific alleles contained within the polymorphisms in the block. This is similar to the idea that all people share the same genes but the spellings of those genes differ, resulting in genetic diversity. Therefore, during crossover when DNA is exchanged between homologous chromosomes, blocks of DNA are exchanged at defined boundaries.

We all appear to share the same large blocks of DNA, known as **haplotype blocks,** though the spellings of those blocks can differ. A particular block could be quite large (e.g., >100,000 bases) and

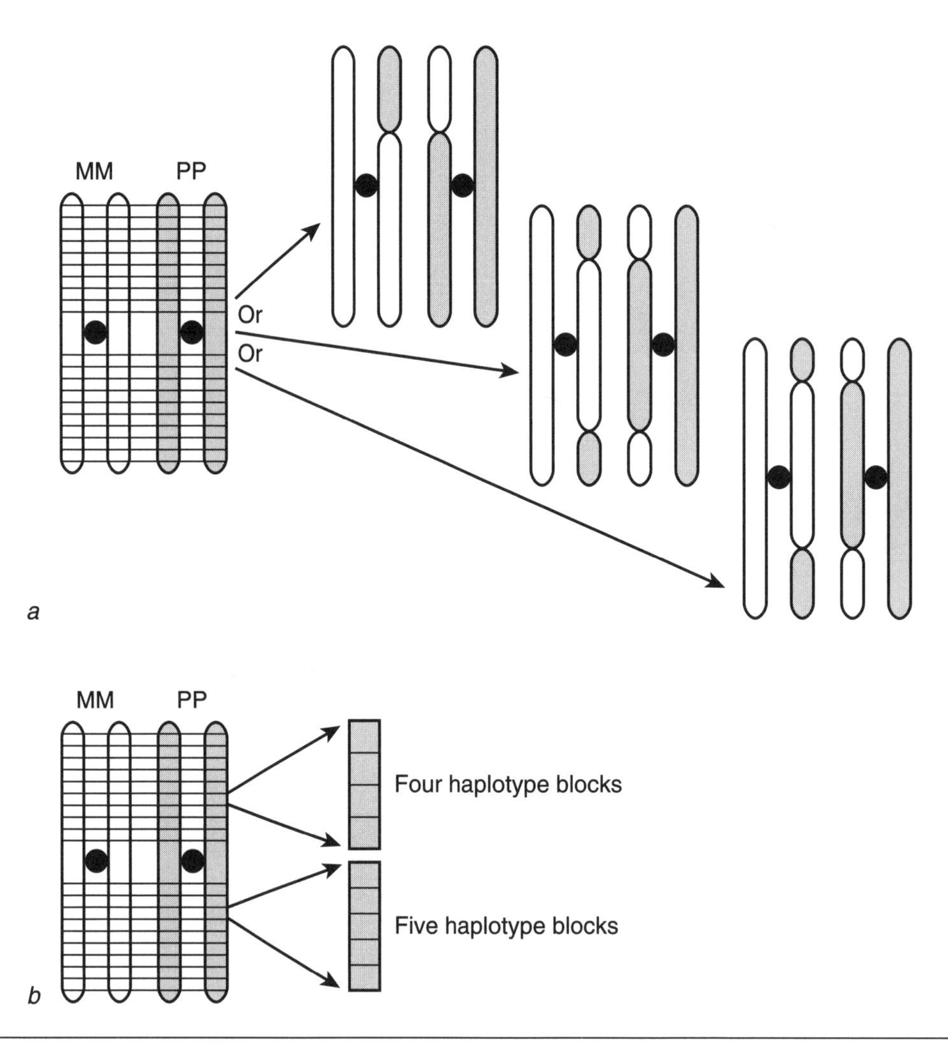

Figure 7.1 While crossover allows the exchange of genetic material between homologous chromosomes (recall figure 4.3), those crossover breakpoints actually occur at specific locations along the chromosomes, known as recombination hotspots. *(a)* Recombination hotspots are shown as lines on the parental chromosomes. During crossover, blocks of DNA sequence can be exchanged at any of those recombination hotspots, meaning that many combinations of DNA exchange are possible (three possibilities are shown; there are many more). *(b)* What is not clear from figure 7.1*a* is that these larger recombination blocks are actually built of many smaller haplotype blocks. In families, this means that DNA is exchanged as large segments of sequence known as *recombination blocks* (each containing many haplotype blocks). Across the human population, hotspot locations can be identified as the boundaries between different haplotype blocks, many of which would be exchanged in a single recombination event as a recombination block.

contain multiple genes, or could be smaller (e.g., < 5,000 bases) and contain only a portion of a single gene. Any single recombination in an individual could result in the exchange of several neighboring haplotype blocks between homologous chromosomes, but recombination events that occur across a population will nearly always occur at haplotype block boundaries shared by the population. In other words, an individual recombination (i.e., crossover) event results not in the exchange of a single haplotype block, but in the exchange of a larger group of neighboring haplotype blocks. Across a population, the many individual crossover events will tend to occur at smaller haplotype block boundaries. Thus, we think of *recombination blocks* when considering individual (or family) recombination events, while we think of *haplotype blocks* when considering recombination events in populations.

The boundaries of haplotype blocks are known as **recombination hotspots,** in that they represent the locations where the DNA sequence is broken in the process of recombination. These hotspots represent the places where crossovers occur: The ends of each block on each homologous chromosome replace each other at the same location, ensuring that the overall structure of the chromosome remains intact while the spelling of the structure becomes unique.

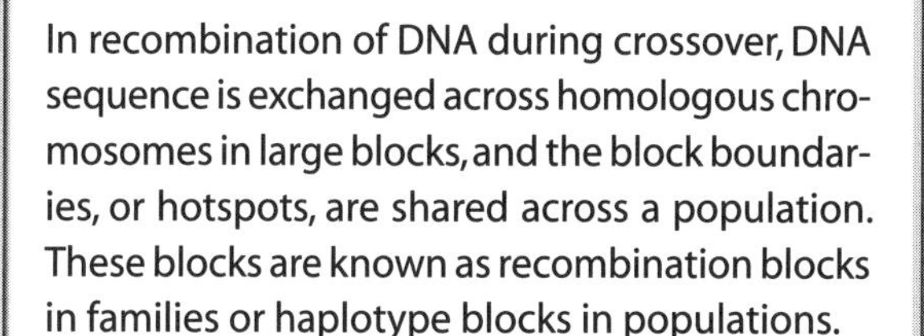
KEY POINT

In recombination of DNA during crossover, DNA sequence is exchanged across homologous chromosomes in large blocks, and the block boundaries, or hotspots, are shared across a population. These blocks are known as recombination blocks in families or haplotype blocks in populations.

There is a unique consequence of the exchange of DNA between chromosomes in the form of haplotype blocks rather than individual letters: The entire segment of DNA within a haplotype block moves together, meaning that all of the letters within a particular haplotype block move together during recombination and are thus "linked." In other words, if you know the sequence of the first half of a block, you can determine the last half of the sequence because you know the boundaries of the block and can identify the block from which the first-half sequence originated. All of this means that the *letters within a haplotype block are predictive for each other:* Knowing some information about the block tells you something about the rest of the block because all of the letters within a block are linked together during recombination. As we'll see in the rest of the chapter, specific alleles within polymorphisms located in a haplotype block are similarly linked, meaning that some alleles tend to travel together during recombination.

LINKAGE DISEQUILIBRIUM

The idea that neighboring nucleotides in DNA sequence tend to travel together during recombination and thus can be predictive for one another is known as **linkage disequilibrium.** This term sounds more complex than it is, so let's break it down to understand its meaning more fully. *Linkage* is simply the idea that two different neighboring nucleotides are linked or connected in some way, which we define as tending to travel together during chromosome recombination (e.g., both nucleotides lie within a single haplotype block). The disequilibrium term requires a bit more work to understand. If we were to take two dice and roll each of them individually, we would predict that the number generated on the first would be completely random and independent from the number on the second at each roll. If we continued rolling in this way, we would reach some *equilibrium,* in which the total numbers of 1s, 2s, 3s, and so on would be roughly equal for the two dice but whatever numbers occurred on the two dice during any particular roll would be random. If, on the other hand, every time we rolled the two dice the same number appeared on both dice, we would easily recognize a nonrandom pattern (i.e., "loaded" dice). In other words, the dice are not in a random equilibrium, but are said to be in *disequilibrium.* To describe two nucleotides that tend to travel together in a nonrandom way during chromosomal recombination, we use the term *linkage disequilibrium.* In other words, the two nucleotides are linked in a nonrandom way, such that one nucleotide travels with the other nucleotide consistently during recombination events. An example is shown in figure 7.2.

When polymorphisms are contained within a haplotype block, the alleles carried by those polymorphisms will tend to be in linkage disequilibrium. Knowing the particular allele at one of the linked polymorphisms, we can predict which other alleles are carried at the other polymorphisms in the block. In other words, one allele can act as a "marker" for any other linked alleles within the haplotype block. This predictive value for alleles within a haplotype block allows for special types of genetic analysis such that more genetic variation can be studied at one time in comparison to what occurs in the typical single-polymorphism studies discussed so far.

Another important point about linkage disequilibrium is relevant to genetic association studies in general: When a positive association has been identified between a polymorphism and a phenotype, it is quite possible that the associated polymorphism is going to be in linkage disequilibrium with other nearby polymorphisms. This is important to recognize, because it is possible that one of these other linked polymorphisms is actually the causal or functional polymorphism in the association. In other words, while we focused our association study on SNP #1 for a particular gene and identified a significant association between SNP #1 and our phenotype, the possibility remains that the true biological source of our association is a different, nearby SNP, one that is in linkage disequilibrium with SNP #1. Thus, very often when a genetic association is reported for a polymorphism that is not known to have functional consequences (e.g., nonsense, missense), researchers will often argue that the polymorphism identified as important may instead be in linkage disequilibrium with the truly functional polymorphisms located nearby.

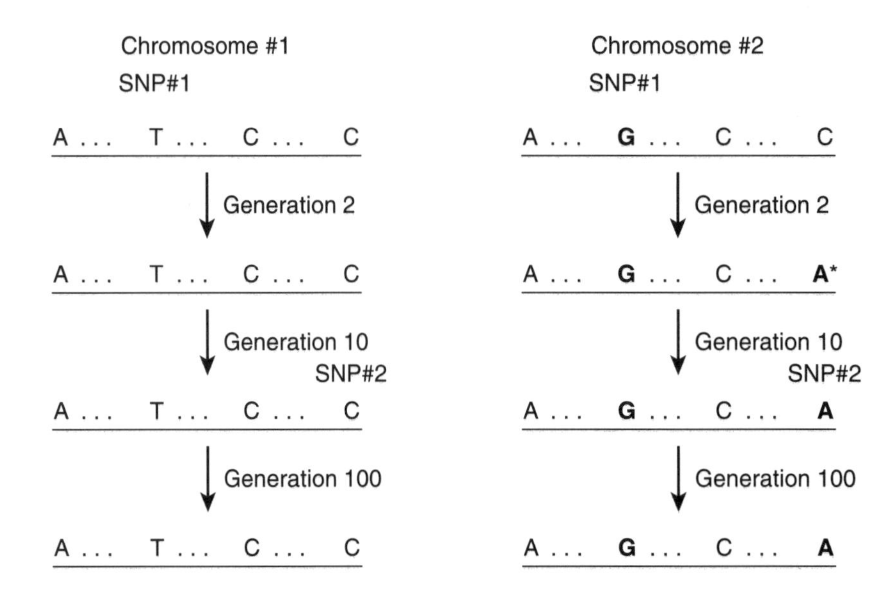

Figure 7.2 Two different DNA sequence segments from a specific region of the same chromosome. This DNA region contains one single-nucleotide polymorphism (SNP) in generation 1, with the chromosome sequence on the left carrying the T allele and the chromosome on the right carrying the G allele. Notice in generation 2 that a novel mutation exists on one of the chromosomes. Over many generations, this is passed along to other members of the population, thus becoming a polymorphism (labeled as SNP #2). If we look at the alleles carried at each of the two polymorphisms, notice that over time the A allele of SNP #2 is always found with the G allele of SNP #1; there is no evidence of recombination occurring between the location of SNP #1 and SNP #2 on the gene. In other words, within the haplotype block that carries SNPs #1 and #2, the alleles T and G tend to travel together over generations, and the alleles G and A tend to travel together—they are not exchanged within the block. Notice also that the other nearby nucleotides are also traveling together; as these are not polymorphisms, there is no variation at the position. Thus, SNP #1 and SNP #2 (or alleles G and A or T and C within those SNPs) are said to be in *linkage disequilibrium*, and knowing one allele of the pair can be used to predict the other allele.

Linkage disequilibrium can be measured for any two alleles on the same chromosome in the genome, the basic pattern being that the closer together the two alleles are physically along the DNA sequence, the greater will be their linkage disequilibrium. Alleles in two nearby polymorphisms, especially those within a single haplotype block, may be in *complete linkage disequilibrium,* such that one allele in the first polymorphism *always* travels with the same allele in the second polymorphism (as shown in figure 7.2). Two polymorphisms that are physically farther apart might be in *partial linkage disequilibrium,* which occurs when specific alleles in two polymorphisms within a block *tend* to travel together, but do not do so 100% of the time. In other words, recombination events occur between the polymorphisms occasionally, such that the two polymorphisms are on opposite sides of the recombination (crossover) breakpoint, but most of the time the linked alleles travel together during crossover.

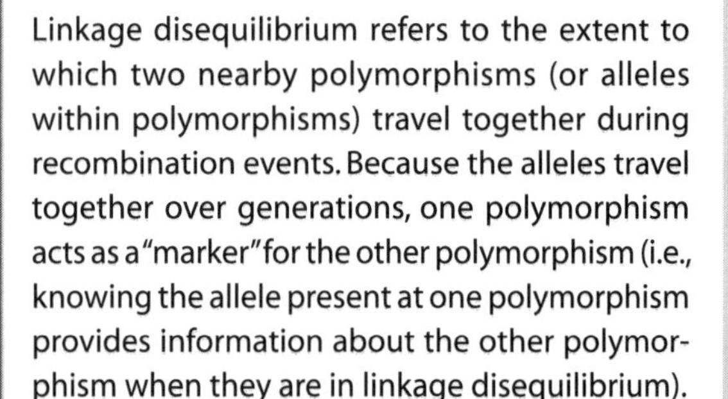

KEY POINT

Linkage disequilibrium refers to the extent to which two nearby polymorphisms (or alleles within polymorphisms) travel together during recombination events. Because the alleles travel together over generations, one polymorphism acts as a "marker" for the other polymorphism (i.e., knowing the allele present at one polymorphism provides information about the other polymorphism when they are in linkage disequilibrium).

The measurement of linkage disequilibrium is beyond the scope of this text, but a minimal introduction to the typical measurements and values is useful. In effect, the measurement of linkage disequilibrium, often abbreviated as LD, is based on the comparison of actual allele frequency data with expected data; in effect we ask how much different the actual data are from what would be expected under random recombination conditions. To return to the dice example, we are testing whether, over many throws, the two dice show random numbers between them or not. The two most common measures for LD are D', also known as Lewontin's D', and r^2. D' is scaled using absolute values from 0 to 1, with 1 representing complete LD. Values less than 1 are hard to interpret, especially for small sample sizes, but at a minimum they represent less than complete LD for two alleles.

The r^2 measure is similar, being scaled from 0 to 1, and is simply a measure of correlation between alleles at different sites. In the case of r^2, however, intermediate values are more interpretable, with higher values indicating greater correlation and thus greater LD between two alleles.

Knowing that alleles can travel together has great utility when one is examining genetic variation in association studies, as we'll see when we discuss the idea of haplotype. First, let's consider how the exchange of the larger recombination blocks in families can be used to examine genetic factors in linkage analysis studies.

LINKAGE ANALYSIS IN FAMILIES

As mentioned in chapter 5, families are the first source of information about the contribution of genetic factors to a trait of interest. Familial aggregation provides our first hints that genetic factors are important for a trait, and additional family-based studies provide the means for estimating heritability—the extent to which genetic (rather than shared environmental) factors are contributing to the trait. Comparing families is also useful for gaining hints about which genes or genomic regions are important contributors to the trait of interest, which can provide an important intermediate step between heritability studies and actual gene identification. This type of study is known as a linkage study or **linkage analysis;** an example will be briefly discussed relative to Alzheimer's disease in chapter 11.

The basis for linkage analysis is that families share large stretches of their DNA because of the process of crossover. As outlined earlier, an individual recombination event will result in the exchange (and thus sharing) of many neighboring haplotype blocks. Similar to alleles in linkage disequilibrium within a haplotype block, alleles within the larger recombination blocks shared by family members are also linked and are thus predictive for other alleles within the larger block. Linkage analysis is performed by genotyping of family members for hundreds of "marker" alleles scattered across the entire genome. Each of those marker alleles will fall within a recombination block, and thus will be predictive for any functional alleles that might be carried in that block. By examining many families and looking for correlations between marker alleles and the phenotype value across those families, one can identify genomic regions as being influential for the phenotype. A correlated genomic region can then be targeted for the identification of candidate genes within that region, thus improving the ability to prioritize the many possible genes underlying a complex trait.

With that overview of the process, let's examine some of the details of family-based linkage analysis. Various study designs are available, including siblings, parent–offspring combinations, and extended family combinations. Different combinations of family members will have different levels of DNA sharing, which can be predicted from Mendelian inheritance patterns. Multiple families are recruited into a study with the goal of identifying genomic regions important to all of the families. Each family will likely have its own unique genetic factors influencing a trait of interest; but with the goal of identifying genes important to entire populations, researchers study multiple, unrelated families to identify genomic regions potentially important at a population level. Families may be recruited because of the presence of a specific phenotype (e.g., disease history, disease risk factor) or could be more randomly recruited, with the phenotype subsequently measured in all family members. In addition to the phenotype measurement, DNA samples are obtained for all family members and are genotyped for several hundred marker alleles spread across the genome. Until recently, 500 to 750 markers were commonly genotyped for linkage analysis studies, but the newer technology (see chapter 10) is now allowing several thousand markers to be genotyped. The greater the density of markers genotyped across the genome, the more specific the linkage analysis can be in pinpointing genome regions of importance.

Once the genotypes are obtained, various statistical procedures are performed to test for correlations among the marker genotypes and the phenotype, taking into account the family relationships and the extent of DNA sharing for different family members. The result is known as a **linkage map,** which is a graphical representation of the extent of correlation for different markers across the various chromosomes. An example of a linkage map is shown in figure 7.3. The figure is made up of several individual graphs, each representing a particular chromosome. The Y-axis of each graph represents the **LOD score,** the statistical metric calculated in linkage analysis that can be thought of as a measure of correlation between the marker allele and the trait of interest. LOD stands for "logarithm of the odds," so this is a logarithmic scale on the Y-axis. When a trait is correlated with a marker allele, the LOD score will be higher. The X-axis represents the length of the chromosome, with the various marker alleles scattered along the length of each chromosome. Thus, researchers look for linkage "peaks," which represent areas of strong correlation on a particular chromosome. Typically, LOD score values greater than 2.0 are considered significant, with LOD scores greater

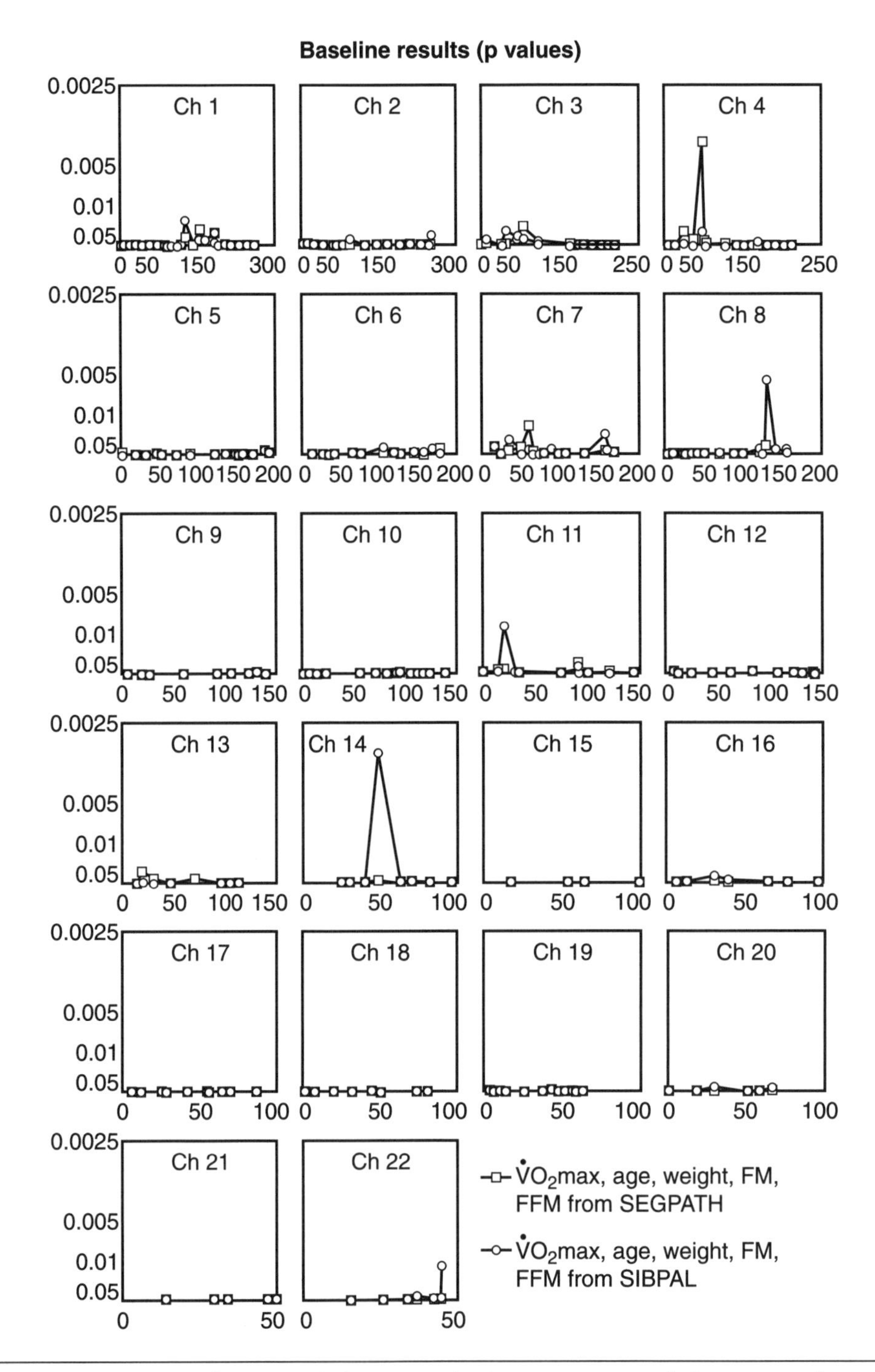

Figure 7.3 A linkage map for baseline $\dot{V}O_2$max. Each chromosome has its own map. For each chromosome, the LOD score (Y-axis) is charted for each of the linkage markers genotyped on that chromosome (in this case, the P-value for each LOD score is shown). When a marker allele is correlated with the phenotype of interest, a higher LOD score will be calculated, resulting in a peak at that region of a particular chromosome. Notice the peaks for chromosomes 4, 8, 11, and 14 on this map. In this example, the high peaks in chromosomes 4 and 14 would be good places to look for candidate genes potentially important for baseline maximal oxygen consumption.

Reprinted from C. Bouchard et al., 2000, "Genomic scan for maximal oxygen uptake and its response to training in the HERITAGE Family Study," *J. Appl. Physiol.* 88(2): 551-559. Used with permission.

than 3.0 generating considerable interest, especially if replicated in multiple studies. The genome region underlying a high LOD score linkage peak would then be a target for the identification of individual genes located in that particular chromosomal region, which could then be analyzed individually in future genetic association studies.

A final note on linkage maps: If few markers are present on a particular chromosome, any identified peaks will be broad and will overlie large stretches of genome sequence, making them less useful for identifying specific candidate genes. Studies with large numbers of markers, especially those using the newer technology, will have much finer-scale peaks, thus improving the ability to identify candidate genes in the underlying genomic regions.

The foundation of all linkage analysis comes down to the presence of linkage disequilibrium between the marker allele and a nearby functional allele (i.e., in the same block). The marker alleles chosen for genotyping in a linkage analysis study are not chosen because they are thought to be functional. Rather, microsatellite repeat polymorphisms are traditionally used as markers, as they have more variation (more possible alleles) than a SNP, providing more detailed genetic information for researchers performing a linkage study. Moreover, these repeat polymorphisms are typically located in noncoding regions of genes or outside of genes altogether, so they are not expected to be functional. Once such a marker allele is considered correlated and important to the phenotype, the location of that marker in the genome will be studied intensely to determine the likely haplotype blocks in the nearby region that might contain a functional allele. This can be a very long and challenging process.

HAPLOTYPE: COMBINATIONS OF NEIGHBORING ALLELES

The consequences of DNA's having a haplotype block structure are profound from the perspective of genetic variation. Just as a polymorphism has multiple alleles, a haplotype block has multiple spellings or allele combinations across individuals. **Haplotypes** are simply combinations of nearby alleles in linkage disequilibrium, such that a set of neighboring alleles travels together during recombination and these alleles are thus informative for each other. An example is shown in figure 7.4. Notice that any particular SNP within a haplotype contains only two alleles; but when several polymorphisms exist in the same block, many unique combinations of alleles are mathematically possible within the block. In fact, the formula for identifying the total number of possible haplotype spellings within a block is 2^n, where n is the number of SNPs within a block. Because most haplotype blocks contain as many as 30 to 70 SNPs, the total possible number of haplotypes can be enormous (even 10 SNPs in a block, 2^{10}, results in 1024 possible haplotypes!). Fortunately for researchers, few of these possible combinations are actually found in any population. For any haplotype block in the genome, there are typically only three to five common haplotypes within that block,

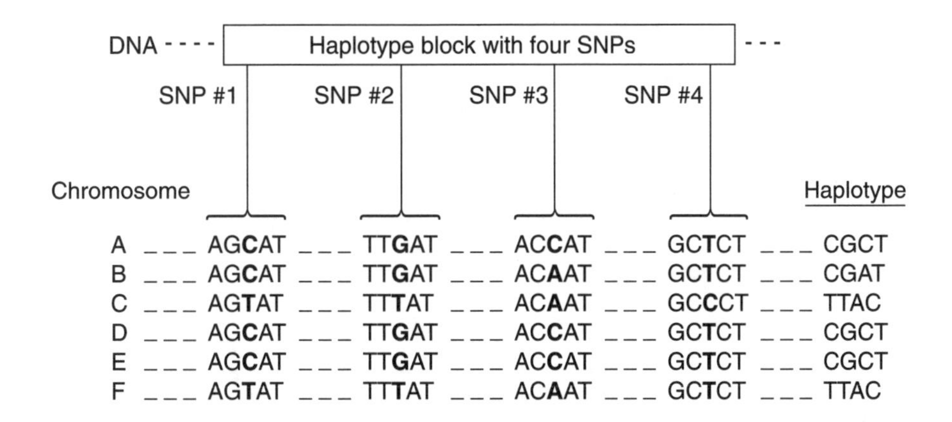

Figure 7.4 A haplotype block in which four SNPs are present. The alleles within these four SNPs are thus in linkage disequilibrium and tend to be carried together during recombination events. The particular pattern of alleles carried within a haplotype block determines the haplotype for that chromosome. In this example, the combination of C, G, C, and T alleles carried for chromosome A at the four SNPs can be called haplotype "CGCT." Similarly, different allele combinations at the four SNPs can result in other haplotype spellings. In this example, six different chromosome sequences were examined, and three different haplotypes were identified: CGCT (carried by three chromosomes), TTAC (carried by two chromosomes), and CGAT (carried by one chromosome).

with "common" defined as being present in at least 5% of individuals in a population. Just as there are very rare alleles present in as few as one person in a population, very rare haplotypes also exist for each haplotype block, but our focus is on those haplotypes that are common in a population and thus suitable for study from a genetic association perspective.

The other unique attribute of alleles within haplotypes is that very often some alleles are in complete linkage disequilibrium with other alleles, making one allele completely informative for one or more other alleles in that block. Thus, when we are genotyping alleles to determine haplotype in an individual, complete linkage disequilibrium makes it unnecessary to genotype all of the 30+ SNPs in the block. In fact, most data suggest that only 10% to 25% of all SNPs in a haplotype block will need to be genotyped in order to accurately define the specific haplotype spelling for that block. The SNPs used to define haplotype are thus known as **tag SNPs,** in that they act as tags or markers for other alleles within the block. An example of tag SNPs is shown in figure 7.5. Researchers are able to genotype this smaller subset of tag SNPs and thus determine the full haplotype spelling for any particular block because of the principle of linkage disequilibrium.

How are haplotypes potentially more important than single polymorphisms in genetic association studies? The simple answer is that more of the genome is being studied when haplotypes versus single polymorphisms are used; thus the opportunity for identifying important genome regions is improved. In fact, the major goal of haplotype-based analysis in genetic association studies is very similar to the goals of linkage analysis in families. In linkage analysis, marker polymorphisms are used to identify large genome regions correlated with the trait of interest in family members, which share large blocks of DNA sequence. In haplotype analysis, tag SNPs are used to identify smaller genome regions associated with the trait of interest in unrelated individuals, which share smaller (haplotype-size) blocks of DNA sequence. The similarity between the two approaches is that the causal allele (i.e., functional allele) is not necessarily expected to be one of the marker polymorphisms that is genotyped; rather the markers will be in linkage disequilibrium with the nearby functional or causal allele. In identifying the smaller block region by relying on linkage disequilibrium,

KEY POINT

Tag SNPs can be used to identify a specific haplotype without the need for all polymorphisms within that haplotype to be genotyped individually. This is possible because some SNPs within a haplotype will be in complete linkage disequilibrium and will be completely redundant for each other.

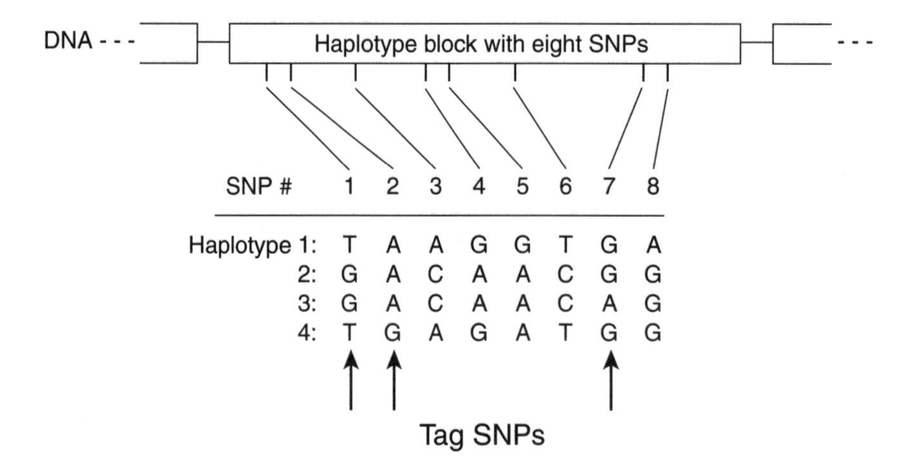

Figure 7.5 A haplotype block with eight SNPs. Across those eight SNPs, four different haplotype spellings are seen in a population (haplotype 1 is most frequent, with haplotypes 2-4 having lower frequencies). Not all eight SNPs must be genotyped in order to determine the particular haplotype carried by an individual. In fact, only three of the SNPs are needed to determine the specific haplotype (indicated by arrows; note that other combinations of SNPs could also be chosen as tag SNPs). These three SNPs are known as *tag SNPs.* For example, notice that the alleles carried in SNP #1 (T and G) match perfectly the alleles carried in SNP #3 (A and C), SNP #4 (G and A), and SNP #6 (T and C), thus making the genotyping of all four redundant and inefficient. Similarly, SNP #5 and SNP #8 are completely redundant for each other. Different combinations of SNPs can be identified as tag SNPs, but the goal is to identify the lowest number of tag SNPs that can identify all haplotype sequences. Various software programs are available to do this.

researchers will have an easier time identifying the actual functional allele because there will be less genome sequence to investigate.

The broader aim of haplotype analysis is the development of *genome-wide association studies*, which are very similar cousins to linkage analysis studies but are performed in unrelated individuals. Just as the name implies, genome-wide association studies rely on the genotyping of haplotype tag SNPs across the entire genome, such that tag SNPs associated with the trait of interest will identify a particular haplotype block region in which to focus subsequent efforts to identify a functional allele. Most estimates suggest that as many as 1 million tag SNPs are necessary to ensure adequate coverage of all of the many haplotype blocks across the genome, although the number may be smaller. As you will learn in chapter 10, currently available technology already allows genotyping of as many as 500,000 SNPs, and similar technology focused on genotyping haplotype tag SNPs is under development. Thus, unrelated individuals will be recruited using standard means for genetic association studies (see chapter 9) and genotyped for hundreds of thousands of tag SNPs across the genome, allowing researchers quicker identification of the key regions of interest for more focused study.

SPECIAL FOCUS
The International HapMap Project

After the Human Genome Project was completed, various analyses of the human genome sequence showed large regions of strong linkage disequilibrium, or haplotype blocks, across the entire genome. As discussed in this chapter, many alleles within the same haplotype block are informative for each other, thus allowing the use of "marker" polymorphisms or tag SNPs for genetic association studies. If tag SNPs could be identified for all of the haplotype blocks across the genome, the ability to perform genetic association studies would be improved as researchers would have the option of studying tag SNPs in linkage disequilibrium with functional alleles, without the requirement of identifying the functional alleles in advance of the study. This rationale was the guiding force behind the assembly of an international consortium of researchers in several countries to form the *International HapMap Project* (figure 7.6).

The HapMap project's primary goal was to extensively study genetic sequence variation across the genome of several race groups, characterizing where gene sequence variants exist and how their frequencies differ among populations from around the world. Men and women from four different populations (African, Chinese, Japanese, and Northern European) were recruited for the first, major phase of the project, namely performing extensive genotyping across the entire genome for the identification of haplotype blocks, tag SNPs, and linkage disequilibrium structure. If there are 10 million or more common SNPs across the genome, the HapMap's basic goal was to identify some smaller fraction of those that could be used as tag SNPs for the many haplotype blocks that exist across all four populations. The inclusion of groups with different geographical ancestry was aimed at identifying possible differences in haplotype block structure and linkage disequilibrium patterns among those groups.

Figure 7.6 The International HapMap Project logo.
From www.hapmap.org.

(continued)

SPECIAL FOCUS

(continued)

Begun in 2002, a first draft of the **HapMap** was published in late 2005, containing more than 1 million SNPs genotyped in 269 individuals across the four populations. An expanded HapMap database of over 3 million SNPs was made available at www.hapmap.org in 2006, and the project researchers will continue by performing more genotyping in the four populations, as well as examining tag SNPs in other populations to confirm their usefulness for genetic association studies. Ultimately, the HapMap gives researchers a tool with which genetic association studies can be performed across larger areas of the genome. Rather than focusing on a small number of polymorphisms in a single gene, researchers will be allowed with use of the tag SNPs to examine (via linkage disequilibrium) much larger segments of genome sequence with fewer genotypes. And with the continued development of high-throughput genotyping methods, the ability to perform genome-wide association studies in unrelated individuals is no longer science fiction.

The HapMap is not without its detractors, who argue that haplotype boundaries are not easily defined and that the size and location of blocks are likely to be different if different tag SNPs are genotyped. Ultimately, even knowing these limitations about using haplotypes for genetic association studies is meaningful, so few doubt that the HapMap will provide important information for human geneticists. In fact, even by the time this section was written, the draft version of the HapMap was used successfully for the identification of a gene polymorphism important for the development of age-related macular degeneration.

DIPLOTYPE: COMBINATIONS OF HAPLOTYPES

Thus far, the use of haplotypes in genetic association analysis has been presented as a fairly straightforward identification of tag SNPs, from which haplotype can be determined and used in genetic studies. Unfortunately, the use of haplotypes in such studies is more complex than the discussion so far has indicated.

Recall that for any gene in the genome (excepting the X and Y chromosomes in men), every individual carries two copies of that gene, one from each parent. Thus, for any single polymorphism, we don't simply report the allele carried in a particular gene sequence (e.g., A or G); we report the combination of alleles present in both gene copies, which we call the *genotype* (e.g., A/A, A/G, or G/G). The same holds true for haplotype. Each gene copy will contain a particular haplotype spelling within a haplotype block, and any individual will thus carry a combination of *two* haplotypes, one from each parent, in his or her genome. The pair of haplotypes that each individual carries for a particular block is known as a **diplotype.** Identifying diplotype in an individual is much more challenging than determining a single-polymorphism genotype, and the genetic analysis is also much more complex.

Why is diplotype so difficult to measure? Figure 7.7 illustrates the problem. Any particular person will carry two gene copies for any autosomal gene, which means that he or she carries two different haplotype sequences for any particular haplotype block. In the three examples shown, we see that the maternal autosome and the paternal autosome each have a unique haplotype, the combination of which is known as a *diplotype* (this is similar to two alleles making up a genotype). Thus, each of these individuals would be considered heterozygous for haplotype. When one performs any typical genotyping assay for a single polymorphism, both gene copies are present in the assay, such that the final result represents the combination of the two alleles, which is then read as a genotype, or combination of the two alleles carried by the individual. In determining haplotype, the same types of data are obtained for each of the tag SNPs genotyped in the haplotype block, resulting in a genotype for each tag SNP. Thus, for each person, you get one genotype for each of the tag SNPs genotyped.

The problem for determining diplotype is that the specific combination of alleles that are carried *on a particular chromosome sequence* cannot typically be determined by genotype alone. In other words, which gene copy (Mom's or Dad's) contains which combination of alleles? Identifying the source chro-

> **KEY POINT**
>
> Because we each carry two copies of each autosome (and females carry two copies of the X chromosome), we actually carry two haplotypes for any particular haplotype block, one on the maternal and one on the paternal chromosome. This combination of haplotypes is known as a diplotype (this is similar to two alleles making up a genotype).

mosome of particular alleles is known as determining the **phase** of the alleles, or knowing whether or not alleles at two or more linked polymorphisms are on the same or different chromosomes.

If we consider only two SNPs, the issue of phase (sometimes called "phase ambiguity") is problematic only for individuals heterozygous at both tag SNPs. If a person is homozygous at either SNP, we know that the same allele is traveling on both chromosomes, whereas double heterozygotes could have either of two haplotype combinations: For the neighboring SNPs A/T and C/G, a person heterozygous at both sites could have haplotypes A-C and T-G, or haplotypes A-G and T-C. The problem is further compounded with the addition of more SNPs, which is often the case when one is genotyping tag SNPs for haplotypes (three to five tag SNPs may be needed to distinguish the major haplotypes in a block). A detailed example of this is shown in figure 7.7. Thus, defining haplotype and diplotype for an individual can be problematic.

There are two main ways to determine diplotype for any individual: (1) a brute force molecular analysis, in which the two gene copies are physically separated and individually sequenced to determine the exact haplotype sequences and diplotype, or (2) estimation of diplotype from the tag SNP genotype data using statistical procedures. The first technique is very labor-intensive and costly, and is not used frequently. Occasionally, for smaller blocks with fewer tag SNPs, easier laboratory methods may allow direct haplotype and diplotype determination. The vast majority of researchers opt for estimation techniques, which rely on statistical procedures that estimate the sequence of the two haplotypes based on the total combination of genotypes present at each tag SNP. Various statistical programs have been developed to perform this estimation, and all rely on using the measures of linkage disequilibrium among the tag SNPs to help guess at which haplotype combinations are more likely across a population, and therefore more likely

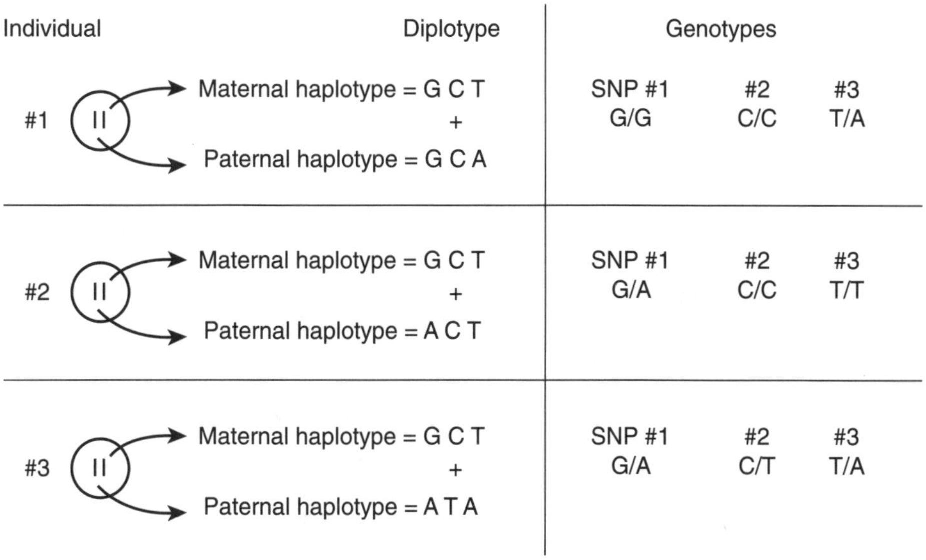

Individual	Diplotype	Genotypes SNP #1	#2	#3
#1	Maternal haplotype = G C T + Paternal haplotype = G C A	G/G	C/C	T/A
#2	Maternal haplotype = G C T + Paternal haplotype = A C T	G/A	C/C	T/T
#3	Maternal haplotype = G C T + Paternal haplotype = A T A	G/A	C/T	T/A

Figure 7.7 A person's diplotype can be challenging to determine directly using only genotype information. If we imagine that each of the three SNPs within the haplotype block was genotyped for each individual (shown on the right side of the figure), diplotype can be predicted for individuals #1 and #2 because they are homozygous for no more than one of the three SNPs. In contrast, the haplotypes and thus diplotype of individuals who are heterozygous for two or more SNPs within a haplotype block cannot be determined without additional information (as is the case for individual #3). In other words, with just the genotypes of G/A, C/T, and T/A for individual #3, we cannot be certain which haplotypes that person is carrying without additional information (i.e., GCT, GTT, GTA, GCA, etc., are all possible).

to occur in any one person. The end result is an educated guess by the **haplotype estimation software** as to the diplotype carried by each individual, with a probability attached to that estimate. Researchers can then exclude individuals who are identified as having less than some threshold level of probability for carrying a particular haplotype or diplotype.

The other issue with diplotypes, assuming that diplotype can be conclusively determined or estimated using statistical procedures, is that, similar to the situation with repeat polymorphisms, any haplotype block has multiple haplotype spellings (or alleles) and thus the number of diplotypes will be considerably larger than the typical three genotype groups seen for a two-allele SNP. So, the statistical analysis is made more complex by the addition of more groups within the analysis, requiring either many more subjects in the study (to increase the number of subjects in each diplotype group) or complex combinations of diplotype groups. Unfortunately, researchers are still working to determine the best ways to perform these types of analyses, so there is little consensus in the field regarding how to go about grouping subjects for haplotype or performing diplotype analyses in general.

This is an area of intense research for statistical geneticists, and the statistical techniques for dealing with haplotypes in genetic association studies are developing and changing rapidly. A recommendation would be to watch for current research literature in your field relying on haplotype-based analyses and use that as a starting point for designing similar analyses within your own studies. These issues are discussed in depth in part II of the book. A more practical recommendation is not to attempt haplotype analyses without the support of a good statistical geneticist!

GENE × ENVIRONMENT INTERACTION REVISITED

This chapter has significantly increased the complexity of the problem of identifying genetic variants important to complex phenotypes. Instead of simply examining one or two SNPs in a gene of interest, we are now considering the examination of many SNPs in a gene region that are in partial linkage disequilibrium, or perhaps the use of tag SNPs across the entire genome to narrow the search for functional alleles. Once specific haplotypes are identified as important, the goal will be to identify the specific functional alleles within or near those blocks, which will then be the focus once again of studies to determine the importance of those alleles on the phenotype in various settings—all of which brings us back to the idea of gene × environment interaction.

Instead of spending the entire chapter on genetic variation and haplotype, I want to reemphasize the fact that genes do not act by themselves but are rather part of a complex system in which environmental stimuli disrupt homeostasis, causing a cell to target certain genes for transcription, resulting in changes in the protein complement of the cell, and thus returning the cell to homeostasis. This is an ongoing cycle in all of the body's tissues, with environmental factors acting on genes and the end result being a cellular response to those environmental factors—the cycle starting anew with every environmental change. And what are these environmental factors? The list is long, but certainly for health and fitness phenotypes the typical factors include physical activity, diet, smoking, medications, and so on. Age and sex can also be considered "chronic" environmental factors (or gene × environment factors), in that the environment of an individual (defined by hormone levels, cellular responses, etc.) is affected by both age and sex. All of these factors can result in changes in gene regulation; and where genetic variation exists within targeted genes (e.g., genotype or diplotype), gene × environment interactions can occur, such that specific genetic profiles will have different cellular responses to the same environmental stimulus. This principle, first outlined in our discussions of single SNPs, is just as meaningful within the context of haplotypes.

AN INTRODUCTION TO EPIGENETICS

Included in the closing of this chapter is a section intimately related to the issue of gene × environment interaction that deals with a topic likely to consume an entire chapter in future editions of this text. Consider this only a brief introduction to an emerging research area.

Recall from earlier discussions in the text that we have considered the variability of a phenotype to be the result of two primary factors: genetic factors and environmental factors. Some combination of those two factors is typically assumed to explain all of the variation seen in a trait, apart from measurement error. Now we are going to add the contribution of a third component to this equation: epigenetic factors. Phenotype variation is thus explained by the varying contributions of genetic, environmental, and epigenetic factors. We have ignored epigenetic factors up to this point in the text because they are a poorly understood contributor to phenotype variation and their full importance is not yet known.

Epigenetics is the study of heritable changes in gene function that occur *independent* of changes in DNA sequence. In other words, information from the parent is transmitted to the offspring, but the information is *not* coded in the letters of the DNA sequence; nevertheless, it influences gene function.

Several examples of such inheritance patterns have been shown in animals and hinted at in humans. For example, hair or coat color in some mouse strains has been shown to be heritable, but independent of genetic sequence variation present in the parents of the offspring. Furthermore, coat color in the offspring of these animals can be altered by the manipulation of the dietary intake of the parents. In humans, a few studies have correlated such phenotypes as birth weight and cardiovascular disease in individuals with the dietary patterns of their grandparents, apparently independent of direct DNA sequence inheritance. Moreover, many identical twins (monozygotic twins) are discordant for the presence of various heritable diseases (i.e., one affected, one unaffected) despite similar shared environments and identical genetic profiles. All of these examples suggest a unique interaction between genes and environmental factors, but gene sequence variation is not part of the equation.

An explanation for these examples lies in the fact that environmental factors can influence DNA, but without changing the nucleotide sequence. DNA can be "decorated" with chemical labels that can affect gene regulation without affecting the DNA sequence. These labels, the most common of which are methyl groups attached to cytosine nucleotides **(DNA methylation),** can be added or removed depending on environmental stimuli presented to the cell. When the methyl group labels are present in the promoter region of a gene or some other regulatory region, they can affect the ability of transcription factors to bind to the gene, thereby affecting gene expression. In fact, for methyl groups, studies have shown that higher concentrations of methyl groups in a gene regulatory region tend to result in lower levels of gene transcription compared to low concentrations of methyl groups. Thus, similar to genetic variants, these chemical labels (sometimes referred to as *epigenetic factors*) can significantly affect gene regulation.

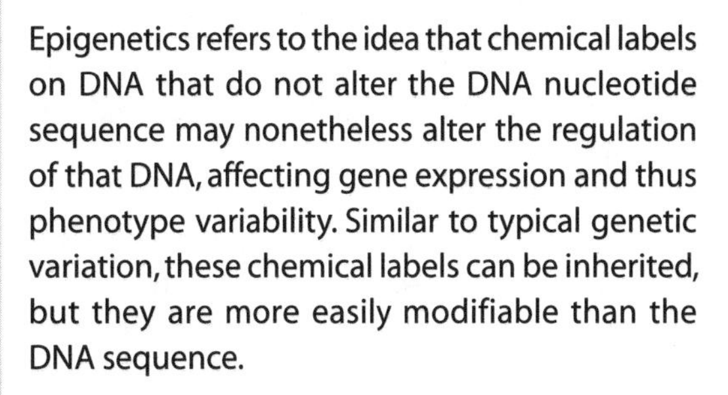
KEY POINT

Epigenetics refers to the idea that chemical labels on DNA that do not alter the DNA nucleotide sequence may nonetheless alter the regulation of that DNA, affecting gene expression and thus phenotype variability. Similar to typical genetic variation, these chemical labels can be inherited, but they are more easily modifiable than the DNA sequence.

The patterns of these labels, called DNA methylation patterns for the case of methyl groups, can be maintained on the DNA during the process of meiosis and can thus be inherited. In this way, methylation patterns important to gene regulation in a parent can be passed along to offspring (and perhaps multiple generations of offspring) without an effect on DNA sequence. This provides an explanation for the examples cited earlier, in which dietary factors affect DNA methylation patterns, which affect gene regulation and phenotypes—and those methylation patterns and phenotype influences are then passed on to subsequent generations. Examples of these influences in humans are only beginning to emerge, but the studies in animal models indicate that epigenetic influences could play a large role in phenotype variability.

So why raise this additional complexity at this point in the text? First, epigenetics is an emerging area that will become more prominent in human genetics research as technologies improve to allow measurement of DNA methylation and other epigenetic DNA modifications. Second, epigenetic factors suggest a novel form of gene × environment interaction that goes beyond the typical examples presented thus far in the text. Up to now, we've basically considered "acute" gene × environment interactions, such that the environmental factor influences gene regulation as the cell responds, and then the response is complete once homeostasis is reestablished. In the case of epigenetics, environmental factors are instead acting in a more chronic way, potentially over multiple generations. I will venture to guess that diet and physical activity (among other environmental factors) will someday be shown to influence DNA methylation patterns in humans, providing an influence on phenotype variability in subsequent generations that is independent of DNA sequence changes.

SUMMARY

The complexity of genetic variation was expanded in this chapter with the concepts of recombination blocks and linkage disequilibrium. Their usefulness in family-based linkage analysis was described, as was the idea of the usefulness of haplotype blocks for association studies in unrelated individuals. Haplotypes are combinations of nearby alleles that are in linkage disequilibrium, and the aim of the International HapMap Project is to fully catalog haplotype structure across the genome, providing advanced tools for use in genetic association studies. Despite the promise of haplotype-based studies, researchers are limited in their ability to measure haplotype combinations, or diplotypes, in individuals, and thus the ability to add these complex variables to genetic association studies. Finally, the idea of gene × environment interaction was revisited, along with the concept of epigenetics, both of which emphasize the close relationship between the environment and genetic regulation, possibly even extending over multiple generations.

KEY TERMS

diplotype
DNA methylation
epigenetics
haplotype
haplotype block
HapMap
linkage analysis
linkage disequilibrium
linkage map
LOD score
phase
recombination block
recombination hotspot
tag SNP

REVIEW QUESTIONS

1. Describe linkage disequilibrium and how it relates to a "marker" allele or SNP.
2. What is a haplotype block? How many haplotype spellings are common in a population for any particular haplotype block?
3. What is a linkage study? How is a linkage study performed, and how is the resulting linkage map interpreted?
4. Describe a tag SNP.
5. What is a diplotype? For the following scenarios, assume we have a three-SNP haplotype of G/C, A/C, and T/G.
    a. What is the diplotype for a person carrying G/G, A/A, and T/G?
    b. What is the diplotype for a person carrying G/C, A/A, and T/T?
    c. What is the diplotype for a person carrying G/G, A/C, and G/T?
6. What are epigenetic factors? How can DNA methylation affect gene function (and a phenotype)?

PART

II

RESEARCH DESIGN AND METHODS

To this point in the book, we have focused on building a foundation of knowledge in the area of genetics, addressing such questions as what is meant by genetic factors, what genetic variation is, and how different alleles can influence a phenotype. With this basic foundation in place, part II explores the application of genetics to health and fitness phenotypes in the real world. Chapters 8 and 9 address the basic issues of identifying genes and polymorphisms for investigation, designing studies, and analyzing data. Chapter 10 follows with an overview of some of the common laboratory methods used to obtain genetic information. By the end of part II, readers should have a strong grasp of the basic issues surrounding the application of genetics in research involving health, physical activity, and sport and will be ready to begin integrating all of that information, which is accomplished with part III.

CHAPTER

8

GETTING STARTED: BASIC SKILLS FOR GENETICS RESEARCH

Whether you are preparing a term research paper on the genetics of $\dot{V}O_2max$ or are considering a study to examine the influence of genetic variants on blood pressure phenotypes, you must address a series of questions in order to understand the existing research literature and properly design a study that will add to that literature. This chapter outlines the first round of questions that are important to deal with at the beginning of any genetics investigation. Using a flowchart, we will approach the key questions to ask when first determining the importance of genetics to a trait of interest. The end result will be the development of specific, testable hypothesis-driven research questions related to a particular phenotype. Here are examples of such questions (many of these are discussed in detail in chapter 11):

- Is the I allele of the angiotensin-converting enzyme (ACE) gene's I/D polymorphism associated with sport performance in endurance events?
- Does the X allele of the R577X polymorphism in the alpha-actinin 3 (ACTN3) gene that results in alpha-actinin 3 protein deficiency influence sport performance in sprint- or endurance-related events?
- Is genetic variation in the ACE gene associated with differences in systolic or diastolic blood pressure? With risk for hypertension or cardiovascular disease?
- Does physical activity alter the increased risk of Alzheimer's disease (i.e., cognitive dysfunction) associated with the ϵ4 allele in the apolipoprotein (ApoE) gene?

IS GENETIC VARIATION IMPORTANT FOR MY TRAIT OF INTEREST?

Figure 8.1 shows a basic flowchart that represents the first questions anyone should ask when preparing to study a genetics question (here we emphasize *human* genetics). The flowchart is a means of addressing the most critical issue before one gets started with the detailed work: Are genetic factors important for my trait of interest? While this may sound a bit too elementary a question, even for a basic textbook, this question lies at the very heart of the problem. If we cannot demonstrate that genetic factors potentially influence our trait of interest, then we have no business beginning the search for candidate genes or performing a research study.

Thus, we begin by defining the phenotype of interest. Remarkably, this can be a bit of a challenge as a first question. If we are interested in cardiovascular disease risk, what does that mean? How will we measure this phenotype? Are we really interested in multiple, related phenotypes? For example, I might be interested in blood lipid profiles or hypertension or C-reactive protein (CRP) levels in the blood—all

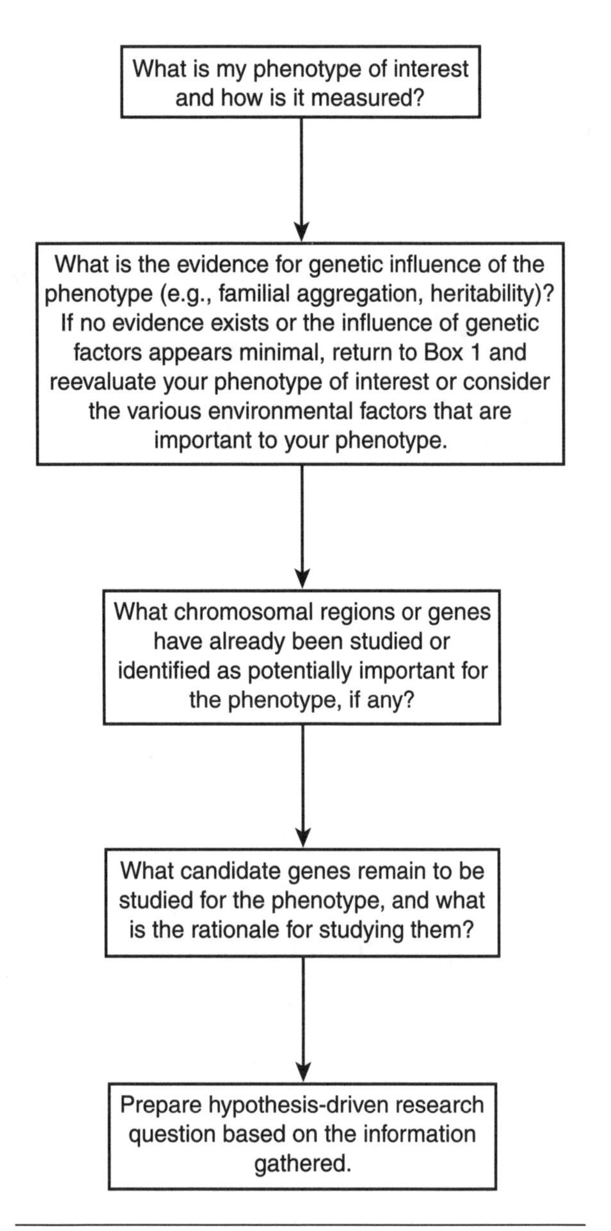

Figure 8.1 The flowchart of typical questions that one should address before beginning a project involving genetics.

of which are important cardiovascular disease risk factors (some as intermediate phenotypes), but all of which could be unique phenotypes in themselves. Through careful definition of the phenotype(s) of interest, the process of determining the importance of genetic factors becomes less challenging.

Once the phenotype is defined, the next basic question to ask is, Has familial resemblance or heritability been reported for my specific phenotype? In other words, have the preliminary studies already been done to suggest that genetic factors somehow influence my trait of interest? This is a key question, for if familial aggregation or heritability has *not* been shown for the phenotype, then there is no support for moving forward with a genetics study! Now, it could simply be that these basic studies have not yet been performed, especially for obscure intermediate phenotypes, so some interpretation of the literature will be required: There is a big difference between the situation in which several studies show little to no heritability for a phenotype and the situation in which no studies have been performed to test for heritability in a specific intermediate phenotype. Thus, if existing research indicates that genetic factors do not play an important role in your phenotype (i.e., there is little to no heritability), then you will want to either reevaluate your phenotype of interest or consider more closely the various environmental factors that are likely important for your phenotype. Alternatively, if no studies exist for your phenotype of interest, that might provide a rationale for performing a twin study as a pilot project to obtain the basic correlation data for determining familial resemblance or heritability (or both), which could then provide evidence that genetic factors are important for the phenotype. Ultimately, as there are few phenotypes of interest to health and fitness professionals that do not have some genetic influence, you are very likely to find studies providing evidence for the importance of genetic factors, and thus the rationale to move to the next phase of the flowchart.

KEY POINT

Defining your phenotype of interest and establishing the importance of genetic factors for that phenotype are the critical first steps in developing a research question.

Once the importance of genetic factors has been established, the next questions revolve again around what has already been done to explain the genetic influences on the trait of interest. For example, have specific chromosomal regions been identified as potentially containing influential genes (such "linkage studies" are discussed in chapter 7)? Have specific genes and gene sequence variants already been studied? Have positive genetic associations already been reported for your phenotype? In other words, has anyone actually begun the process of moving beyond the heritability analysis to start defining specific genetic variants that influence the trait of interest? If not, how will *you* begin to study genetic factors for your trait? Or, if so, how will you *add to or improve* the existing literature in a meaningful way? The remaining sections of this chapter discuss this process in detail.

USING PUBMED

The questions outlined in the figure 8.1 flowchart rely to a great extent on the existing research literature. I will assume that you have some familiarity with the basics of research articles and how to go about finding them, so this section is about defining this process in the context of genetics.

The online search tool known as **PubMed,** operated by the National Library of Medicine, is an incredible resource for identifying research articles published since 1966. This is very often the first stop in any attempt to locate research articles in any medicine-related field (and medicine is defined *very* broadly for inclusion in PubMed!). So, this will be the first stop in any search for familial aggregation and heritability papers, as well as for any gene-specific studies that have already been performed for the phenotype of interest. One caveat is that many heritability studies were performed *before* 1966 for many anthropometric traits, so a negative search result in PubMed should not be defined as conclusive.

So how should PubMed be used for locating articles with a genetic focus? The purpose of this section is really to provide some strategies for narrowing searches, with the keywords most typically used to define genetics studies. For any PubMed search, you will first want to know the key phrases used to describe your phenotype: Is one phrase universally used to describe the phenotype (e.g., cardiac output), or are alternate phrases (e.g., $\dot{V}O_2$max or maximal oxygen consumption or maximal oxygen uptake) used? Has the terminology underlying your phenotype changed over the years, such that older studies may have used different phrases to refer to the phenotype (this is especially important for finding older heritability studies)? Should related phenotypes be considered? The phrases that you define as important for your phenotype will necessarily be included in any of the PubMed searches you perform, in order to limit your search to articles important only to that phenotype.

Once the descriptive phrases have been defined (which may require some work in itself), the search is on to identify the specific articles of interest. For heritability studies, the search words or phrases to add to your phenotype phrases include *heredity, heritability, familial aggregation, familial resemblance,* and *familial similarity.* One or more of these phrases, when combined with your phenotype phrases, is likely to bring up any heritability-type studies performed since 1966.

If you are searching instead for specific studies related to linkage analysis, specific genes, or genetic associations, the following words or phrases will assist with your search: *genetic linkage, linkage analysis, gene, genetics, genetic association, genotype, allele, mutation,* and *polymorphism.* By searching several of these phrases, you will quickly get a sense of the extent of the research literature for your phenotype of interest. If studies haven't been performed yet, then negative or nonspecific search results will be returned. If studies have been performed, you will likely see genetic association studies that have targeted specific candidate genes and polymorphisms, providing a preliminary list of the genes and gene variants already considered important by other investigators. When studies are available, rely also on the reference lists in those papers to screen for other research articles not identified in your PubMed search, especially older studies.

INTERPRETING EXISTING LITERATURE

As will be seen in subsequent sections and chapters in the text, the existing literature identified in PubMed will require some interpretation in order for you to identify the limitations of the existing research. These issues will become more apparent as they are discussed in detail later in the book, but it is important here to state that the existing literature will require careful consideration. While PubMed is a fabulous resource for finding articles under the various search terms just outlined, it does little to tell you how or why these studies were performed. You can address these questions, which are often very important for determining the rationale behind various study designs or candidate genes, only by reading the various research articles and interpreting their findings.

KEY POINT

Use PubMed to identify the existing literature for your phenotype of interest. By interpreting the rationale and findings of those existing studies, you will be better able to understand the current state of the science and develop novel research questions to improve the existing science.

When reading the research literature, focus on a few key issues. What was the specific phenotype under investigation? How was the phenotype measured? Ultimately, is the phenotype, as defined and measured in a particular study, similar enough to your phenotype of interest to warrant inclusion in your review of literature? What subjects were studied? Were they randomly selected, or were they specifically selected based on their health or disease status or some other attribute (e.g., age, sex, race or ethnicity)? How would the selection of the subjects influence the phenotype under study, and does that have relevance to whether the study is informative for your future investigation of the phenotype? Finally, what genes and gene polymorphisms or mutations were studied? The title and abstract may mention only those genes and polymorphisms positively associated with the phenotype, but the results and discussion sections of the research article may detail other genes or polymorphisms that were studied but found to have no association with the trait of interest. What was the rationale used for the selection of the genes and polymorphisms chosen by the investigators, and does that rationale make sense for your phenotype? Does the investigators' rationale tell you something about the physiology underlying the trait of interest that can be informative for your selection of genes to investigate? All of these questions will provide information that will help you understand the current state of knowledge for your phenotype, and will likely also influence the study that you will design to add to the literature.

IDENTIFYING AND SELECTING CANDIDATE GENES AND POLYMORPHISMS

Whether specific genes have been previously studied for your phenotype or not, the next steps for moving forward with a research study are to (1) identify potential **candidate genes,** (2) prioritize those genes and select the genes you will study from that list, and (3) identify polymorphisms within those genes that will be investigated in your study. Besides the existing literature for your phenotype, the most important source of information for this process is the known physiology underlying that phenotype. In other words, by understanding the regulatory systems and structures that are important to the function of the trait of interest, you can identify the candidate genes that are likely to be important. The proteins that are critical to these various regulatory and functional systems are coded for by genes, and those genes become the candidate genes of interest for your investigation.

This rationale for candidate gene selection will likely be obvious in any existing literature for your phenotype: The genes already studied are likely to code for proteins already known to be important for the function of the phenotype. If you decide to study genes and polymorphisms that have already been reported in the literature, then you will want to make sure you are improving that literature by designing a question that expands or improves on the existing research. On the other hand, if you choose a gene that has not yet been studied, the physiological rationale for the candidate gene you select for study will need to be strong in order for others in the field to accept your research as important.

Unless you have identified a very specific intermediate phenotype, the odds are good that you will be able to generate a very long list of potential candidate genes of interest because the physiology underlying most health- and fitness-related phenotypes is complex and is affected by multiple regulatory systems. So the next step is to prioritize those genes: Make an educated guess based on your knowledge of the phenotype about which genes are most likely to be important. The following are a few strategies for doing this.

The first way to move genes up on the priority list is by recognizing the importance of their proteins within the research literature. What genes on your list have been most intensely studied over the past few years as being very important to the underlying physiology of your trait? Have animal models been established for your trait of interest, and have genetic manipulations in those models identified important genes? What physiological pathways underlying your trait have the least redundancy (i.e., are most important to the regulation of your trait), and what are the key proteins (i.e., genes) within those pathways? The answers to these questions will be found in an extensive reading of the basic research articles for your trait of interest. Very likely, you will find that only a small fraction of your very long list of genes has been the target of intense investigation, providing a reasonable means of focusing your candidate gene list. This is not to say, however, that these will always end up being the most important genes underlying the genetic influence on your trait of interest; if we had that information in advance, there would be no need for this prioritization process! The point is that the genes identified as critical in basic physiology are important first genes to examine from a genetic variation perspective and should be considered first.

A second idea to consider is the specificity of the expression of the genes on your prioritized list. Is the candidate gene expressed only within the tissues of interest for your phenotype, or is the gene expressed widely in many cell and tissue types in the body? If the gene is ubiquitously expressed (i.e., transcribed in many cell types), the odds are less likely that genetic variation in that gene will have a large impact on the gene's function, because a large number of cell and tissue types would be affected by such genetic variation. Thus, genes that are ubiquitously expressed should be moved down the priority list, despite their potentially critical importance for the trait of interest. Conversely, those genes that are expressed only in the tissue of interest should be considered as a higher priority, as genetic variation in those genes may affect the tissue of interest without consequences in other tissues.

KEY POINT

Rely on the underlying physiology of your phenotype of interest to identify candidate genes and develop the rationale needed to propose those genes for a new research study.

At this stage, you've probably generated a fairly condensed list of candidate genes to consider for further investigation. The next step with respect to this narrowed list is to identify the genetic variation (i.e., polymorphisms) within the genes. Which genes have polymorphisms most likely to influence the function of the gene or protein? Recall the discussion in chapter 5 about how certain polymorphisms, such as nonsense and missense single-nucleotide polymorphisms (SNPs), are far more likely to influence protein function than others, such as intronic SNPs. Which of the candidate genes on your list have high-priority polymorphisms that are most likely to affect gene function? Ultimately, the goal of genetic association studies is to identify those genes and alleles with the greatest influence on interindividual variation in the phenotype. If a gene is important to a phenotype but does not have polymorphisms that are likely to influence the function of the gene or protein, the odds are not strong for that gene being a major contributor to phenotypic variation. Thus, careful consideration will need to be given to the polymorphisms within each of the candidate genes of interest.

Table 8.1 shows some of the evidence that a researcher might collect as he or she moves to develop a new research question for a particular phenotype. For several candidate genes identified (each with a strong physiological rationale), each gene will have a certain level of literature support and different types of genetic variation. In table 8.1, genes 1 and 3 appear of interest given consistent literature support (gene 1) and influential polymorphisms (genes 1 and 3). Genes 2 and 4, on the other hand, have less literature support (gene 2) or have polymorphisms of less interest (gene 4). Ultimately, the researcher will need to make the final decision based on the sum total of all this information. Does it make the most sense to pursue a previously studied gene in a new way to improve existing literature, or was a completely novel candidate gene identified that has a particularly interesting polymorphism (e.g., nonsense or nonconservative missense SNP)? The answers to these questions will vary in different situations, but very often the typical constraints of developing and executing a research study (as discussed in chapter 9) will help push you toward a particular gene on your list.

Table 8.1 Example of Evidence Collected in the Development of a Research Question for a Genetic Association Study

Candidate gene	Polymorphisms of interest	Previously studied for phenotype of interest?	Key results
Gene #1	2 missense (R57G, S44Y), 3 intronic SNPs	Yes	R57G: G allele positively associated with the phenotype in two studies; S44Y not studied
Gene #2	1 missense (N123K) SNP	Yes	N123K: K allele positively associated with the phenotype in one of two studies
Gene #3	1 nonsense (C456X), 1 promoter, 2 intronic SNPs	No	–
Gene #4	1 intronic SNP	No	–

An additional question that will need to be addressed is the number of genes to investigate. Although single-gene, single-polymorphism studies have yielded some important results for various phenotypes, ultimately, multiple genes and multiple polymorphisms will influence heritable complex traits. Moreover, the ability to genotype many polymorphisms in many individuals, as well as the desire to understand more completely the influence of multiple polymorphisms on a trait of interest, has moved many researchers to consider multiple genes and multiple polymorphisms in a single study. The statistical issues underlying such an analysis are beyond the scope of an introductory text, but certainly there are considerable challenges to the statistical analysis of many polymorphisms simultaneously. One strategy that has been considered is to study all of the candidate genes within a particular regulatory pathway for a phenotype. In other words, are a few of the candidate genes on your list within the same pathway, such that they are known to interact with each other? Some have argued that this is a better approach to addressing complex genetic influence than including less-related genes in a single analysis. Even if a single gene is selected as a target for analysis, the polymorphisms within that gene will need to be selected as well. Certainly, focusing on those polymorphisms anticipated to alter gene or protein regulation (e.g., missense SNPs) should be a part of any analysis. But for genes without such high-priority polymorphisms, which lower-priority polymorphisms should be selected and where in the gene? In this case, multiple polymorphisms at locations across the gene should be considered. This issue is addressed in depth in chapter 7, where the issues of linkage disequilibrium and haplotype are discussed.

To summarize briefly the issue of candidate genes and polymorphisms, an important consideration is to be as focused as possible on the goal of identifying in advance those genes with the greatest likelihood of influencing the trait of interest. While this statement is obvious, the details for how to perform this guesswork are complex and will ultimately be specific to the phenotype you are interested in studying. By focusing on the most important proteins (and their genes) in the physiological pathways and, among that limited list, only those genes with polymorphisms likely to influence gene or protein function, you are not only in a stronger position to identify influential and clinically significant genes and alleles but also more likely to convince your colleagues that your rationale for performing such work is sound. As noted previously, the issue of haplotype, which adds further insight (and complexity) to the question of polymorphism selection, is addressed in chapter 7.

USING GENOME DATABASES

So, how do you go about finding polymorphisms and other information for the proteins or genes on your candidate list? During the process of identifying candidate genes, we often focus on the proteins involved in the physiology underlying the phenotype, as these are the actual functional units within the cell. For a genetic study, we require the information for the *gene* that codes for the given protein. Several online databases are freely available for finding this information, and they are discussed briefly in this section.

One of the challenges of candidate gene and polymorphism identification is that the names commonly used to describe various proteins in the body are often different from the names and labels used for the genes that code for those proteins. For example, the gene that codes for the skeletal muscle protein, myostatin, is officially called GDF8, for growth and differentiation factor 8, the original name for the myostatin protein. Thus, once candidate proteins have been identified, it is not always a simple task to identify gene structure and polymorphism information for those genes. In this section, we will discuss strategies for searching the various Internet-based genome databases.

The National Center for Biotechnology Information, or NCBI, part of the National Library of Medicine, developed one of the first comprehensive online databases of biological information, including such things as protein sequences, mRNA sequences, whole genome sequences, gene locations and structures, and, more recently, polymorphism locations and characteristics. These NCBI **genome databases** can be accessed through the main NCBI Web site: *www.ncbi.nlm.nih.gov.* The NCBI Web site has a pull-down search menu of the many databases that can be accessed to specify the type of search you are performing (e.g., protein, gene); even PubMed can be accessed through this main Web site.

Other databases that have been developed with a focus around the human genome sequence contain similar information, and each has its own characteristics. Two commonly used databases are www.ensembl.org, developed by European research groups, and genome.ucsc.edu, developed by the University of California Santa Cruz Genome Bioinformatics group. All three of these databases contain very similar bits of information, though you'll often find that one of the three is more useful for finding particular types of information.

So how should these databases be used once a candidate gene list has been reduced to the highest-priority candidates? Because the candidate gene list will consist primarily of proteins known to be important to the phenotype of interest, the first task is to *identify the official gene name and symbol for each of those proteins.* A useful starting point for finding these connections is the search bar located on the NCBI home page, which screens all NCBI databases using the NCBI Entrez cross-database search engine. Within this main search bar, search for the protein name that you have and look at the search results from the Entrez cross-database search engine. From here, you can select the Gene, SNP, Protein, Nucleotide, and other specific databases to find the information that you need. In the case of moving from protein name to gene name, the Gene database is a useful first choice. Be aware that these databases contain information for many species, including *homo sapiens;* on most sites, you can limit your searches to only *homo sapiens* (or any other species), and this can help eliminate extraneous information.

In the search for an official gene name and symbol from a literature-based protein name, there are three typical scenarios seen in the Gene database. (1) The protein name search brings up a few listings, the first of which contains your protein name, and the protein name is identical to that of the official gene name. (2) The protein name search results in several listings, one of the first of which contains your protein name, but clicking on that entry shows that the protein name is an alternative name of the gene, with the official gene name being different. (3) The protein name search brings up listings that do not show the protein name you searched for, in which case you will be forced to click through some of the entries NCBI presents in order to identify possible alternative names. Alternatively, you may need to return to PubMed and the research literature to search for other names for the protein and restart your search. Ultimately, the goal is to identify the official *HGNC gene name and symbol* for your gene (named by the Human Genome Organization Gene Nomenclature Committee) and verify that it codes for the protein that you are targeting on your candidate gene list. Once a search result has been clicked, it will open up a screen specific to the given gene that contains basic information about the function of the protein. By reading this summary description of the coded protein, you should be able to verify that you've identified the correct gene and can be confident that you've identified the true official gene name and symbol.

Identification of the HGNC gene symbol is much like obtaining a master key to a building: Once you have found it, you have quick access to the many other databases that are available. Because the other genome databases also rely on the same master genome sequence information and HGNC nomenclature, the official gene symbol can be used to search for genome information quickly across all of these databases.

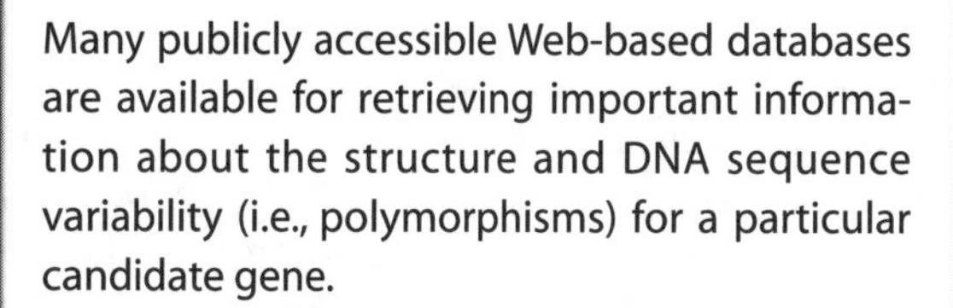

KEY POINT

Many publicly accessible Web-based databases are available for retrieving important information about the structure and DNA sequence variability (i.e., polymorphisms) for a particular candidate gene.

If you have already accessed these various databases, you may already feel comfortable with them and may have preferred databases for various searches. If you are new to these databases, I would recommend identifying the HGNC symbol within the NCBI databases, and then using that HGNC symbol to find specific gene information within the Ensembl database, which I consider the easiest to access for the information typically needed for genetic association projects. When you are searching within the Ensembl database, the HGNC symbol is listed specifically for each search result, allowing you to quickly verify that you've identified the correct gene. The Ensembl database then allows fairly easy access to gene structure, gene sequence, mRNA sequence, protein sequence, and genetic variation information for that gene, though the polymorphism information is not always as complete as that found in the NCBI SNP database.

In the days before the human genome sequence was readily available online, the location and structure of a particular gene, if known, was often critical so that specific DNA sequencing could be performed to identify novel gene polymorphisms in a population of interest. Today, however, the human genome sequence is completely available in these online databases, and the bulk of polymorphisms have been identified and added to the SNP databases. Thus, the primary task for researchers today, once they have identified the official gene name and symbol for their high-priority candidate genes, is to find the polymorphisms that are known for those genes. When you use the genome databases, especially Ensembl, a separate page for each gene will detail the polymorphisms identified for that gene and provide a quick summary of the most important information, including where the polymorphism is, what kind it is (SNP, insertion/deletion, etc.), and whether or not it alters the protein sequence. Hopefully in the near future these databases will also highlight those polymorphisms known to influence gene or protein function, as these will be the highest-priority targets for gene association research. This information is obtained for each of the candidate genes and used to help further prioritize those genes as already described.

Another approach to consider for polymorphism identification is to return to PubMed. Whether or not a gene has been studied in relation to your specific phenotype, the possibility exists that genetic variation in your gene of interest has been studied for a different phenotype and that other researchers have already highlighted what they consider to be important polymorphisms within that gene. You would find this out by searching in PubMed for the protein and gene name or symbol with such search terms as *polymorphism, allele,* and *genotype.* Certainly, if other researchers in the field have already studied polymorphisms within a gene, those variants should be considered for analysis in your study. By including previously published polymorphisms in your analysis, you can perform a replication of the previous work, providing further evidence for or against the importance of that gene for the phenotype. Adding other polymorphisms to those previously studied would be important, however, as older studies may not have included more recently identified high-priority polymorphisms in the gene.

The bottom line when you are using the genome databases is that they are tools to assist with decision making. You'll need to use the criteria described in this chapter and in chapter 9 for choosing among candidate genes and polymorphisms and ultimately deciding whether or not you have the scientific rationale needed to move forward with a gene association study.

SUMMARY

At this point, you've performed the basic tasks needed for developing a review of the genetic literature for your phenotype of interest, or for beginning the design phase of a future genetic association study. The importance of genetic factors for the trait of interest has been determined, and the key literature has been interpreted. Candidate genes from both the literature and direct knowledge of the physiology underlying the phenotype have been identified, and that listing has been further refined to develop a smaller listing of the highest-priority candidates. Those candidates, often known only as proteins at this stage, are then paired with their official gene name and symbol, allowing access to the gene-specific information for each candidate. Finally, the polymorphism information has been secured for each of the candidates; this will likely further refine your candidate gene list depending on the characteristics of the gene variants in each gene. At this point, you are ready to ask another series of questions related to the design of a potential study, which will be addressed in the next chapter.

KEY TERMS

candidate gene

genome database

PubMed

REVIEW QUESTIONS

1. What is the primary question to answer before beginning a project involving genetics?
2. What are the search terms used to identify previously studied genes and gene polymorphisms in the PubMed database?
3. What are the strategies for determining candidate proteins and their genes for a phenotype of interest?
4. Once a candidate protein has been identified, what are the steps for identifying the official gene name and symbol for that protein?

CHAPTER

9

ISSUES IN STUDY DESIGN AND ANALYSIS

This chapter tackles the questions surrounding how to perform genetics investigations for health and fitness phenotypes. It may be surprising to see the number of impacts the addition of a few polymorphisms can have on the design of a study and the analysis of its resulting data. This chapter assumes that the reader has some background in the general aspects of research design and emphasizes only the tweaks on those skills necessitated by the addition of genetics to the study.

HOW GENETICS NECESSARILY ALTERS STUDY DESIGN

Including "genetic variation" as an independent variable in a research study adds a unique context to study design in health, physical activity, and sport research. Suddenly, instead of individual subjects being grouped simply by sex, age, disease status, exercise training status, and so on, subjects are grouped according to genotype. Moreover, for the complex traits that are the typical focus of such studies, the emphasis is now on the influence of genetic factors; and there are often many genetic factors, each with potentially only a small influence on the trait of interest. These slight changes end up having a large impact on the design and execution of genetic association studies. One of the most visible impacts is in sample size. Because the focus is on studying gene polymorphisms that likely have only a small to modest influence on the phenotype, larger sample sizes are needed in order to provide the statistical power for confident identification of true associations. As will be seen in the examples in chapter 11, sample sizes for genetic association studies typically range from 100 to sometimes thousands of subjects. This is in stark contrast to typical research studies on the influence of various environmental stimuli on a phenotype, in which sample sizes of less than 50 are not uncommon. Recall from our discussion of the pie chart in chapter 5 that most complex phenotypes will be influenced by many genes, each with a relatively small direct influence. When we assess the impact of an environmental factor (e.g., exercise training) on a phenotype of interest (e.g., $\dot{V}O_2max$), we often expect to see a substantial effect, for example changing a phenotype characteristic by 10% to 30% (or more). On the other hand, when we are trying to determine the influence of a single gene on a trait of interest, we are often likely to find only a small (1-5%) influence. Such smaller effects are harder to observe statistically without a larger sample size.

Because the emphasis has now shifted from the environmental factors side of the phenotype pie chart to the genetic factors side (recall the pie chart discussion from chapter 5), researchers need to account for as many of the environmental factors affecting the phenotype as possible. Think of any research study as simply working to understand why variability exists in a phenotype across many individuals. Why is $\dot{V}O_2max$ or muscle strength or cognitive function different across a random cross section of a population? Over time, the influences of various environmental factors (e.g., physical activity, diet, disease) are shown to account for some fraction of the phenotype variability observed among individuals. Before the mid-1990s, few studies tackled the influence of specific genetic factors in their study designs because of technological limitations; they simply ignored the genetics side of the phenotype pie chart. When the focus shifts to genetics, the measurement of environmental factors known to influence the trait of interest suddenly becomes even more important. Accounting for all of the environmental factors known to influence the trait of interest makes it more likely that the phenotype differences observed among individuals will be based on genetic differences, and the identification of those factors is the aim of a genetic association study. In other words, accounting for more and more of the environmental slices of the phenotype pie improves the ability to identify slices (i.e., genes and their alleles) on the genetic factors side of the pie.

The following equation is another way to think about the various contributions of genetic and environmental factors on a phenotype:

$$Y_x = m_x \pm G_x \pm E_x \pm (GE)_x \pm \epsilon_x$$

where

Y = value of the phenotype of interest,

x = phenotype of interest,

m = population mean,

G = genetic factors (genotype or haplotype),

E = environmental factors, and

ϵ = error.

Let's work through this equation conceptually to make it clear. This equation is written for an individual, as opposed to a group of individuals, so let's assume that the individual in question is you. For any phenotype of interest, say muscle strength, you will have a certain phenotype value (determined by a muscle strength test). This is true for any phenotype. This is represented in the equation by the term Y_x, where Y is your phenotype value and x simply tells us the phenotype of interest (because this equation is true for any complex phenotype). So, why do you have the muscle strength measurement that you have? Why isn't your measurement value the same as that of everyone else in the population? What are the factors that have resulted in your specific measurement, which make you different from many of the other individuals on the planet? The equation lays this out. Your phenotype value (Y) can be thought of in relation to the population average or mean value (m). In other words, your phenotype value *is* the population mean, but it is altered by the unique genetic (G) and environmental (E) factors that influence your body. For example, your genetic profile will be unique from those of all other individuals in the population (except your identical twin); so if genetic factors are important for muscle strength, you bring to your muscle strength measurement specific alleles that will increase or decrease your measurement value compared to the population mean.

In addition, various environmental factors known to influence muscle strength will also contribute to your value. For example, if you are an older woman who performs little physical activity, your strength value is likely to be lower than that of a young male athlete. That combination of genetic and environmental factors that is unique to you will "change" your phenotype measurement from that of the population mean to a value unique to you. The final terms in the equation are simply combinations of genetic and environmental factors (GE), or what we think of as gene × environment interactions, and measurement error (ϵ), which is an unfortunate source of phenotype variation among individuals (no measurement device or technician is perfectly accurate 100% of the time). The additional term of gene × gene interaction (GG) could be included, but it is considered inherent in the G variable of the equation. Age and sex can be thought of as either GE or GG factors, as they are really combinations of influences governed primarily by genetic factors. From a study design perspective, however, these are "controllable" traits in that they can be included within exclusion and inclusion criteria or data analysis (or both) and can thus be manipulated as part of the experimental design.

So a critical goal of designing a genetic association study can be thought of as controlling for the "E" and "GE" in the equation: controlling for all of the known environmental factors influencing the phenotype value, such that the unknown genetic factors are all that are left to explain differences among many individuals. This means that genetics is truly an addition to all of the other planning that has traditionally gone into careful study design. For example, if we sought to determine why muscle strength differed among individuals taking a dietary supplement, we would also want to assess their age and sex (both GE factors), muscle mass, physical activity levels, and so on before moving forward with our study. The same holds true for genetic association studies.

We finish this section with a final word on phenotype measurement. That "ϵ" term in the equation

KEY POINT

The success of a genetic association study often rests on the researcher's ability to control factors that are known to be important to the phenotype. Controlling such variables as age, sex, exercise training history, diet, and so on is more likely to make the influence of genetic factors on a trait observable.

should not be treated lightly. While measurement error is a part of all research studies, the more accurate the measurement is for any individual, with precision across all individuals in the study, the more likely you will be to minimize the error in the measurement term. And because that ϵ is another contributor to phenotype variability (though variability not due to physiology, in this case), the smaller it is in the equation, the more likely any phenotype differences among subjects in a study will be due to genetic factors. Careful phenotype measurement is critical in studies of genetic association.

TYPICAL STUDY DESIGNS

When one is considering a genetic association study, there are two typical design options: (1) identify phenotype differences among different genotype groups, or (2) identify differences in genotype or allele frequencies between different phenotype groups. Let's clarify just what these scenarios mean in real-life terms.

In the first scenario, a researcher is interested in identifying an association between one or more specific gene polymorphisms and a phenotype of interest. In this scenario, the underlying question is whether or not the phenotype measurement differs between carriers of one genotype versus carriers of another genotype, with the hypothesis that genetic variation contributes to phenotype variability. In other words, for an A/T polymorphism, do A/A homozygotes have significantly different phenotype values compared to T/T homozygotes? We will refer to this as an **association study design,** and examples are shown in figure 9.1. In this design, subjects are recruited regardless of phenotype value (in other words, all levels of the phenotype are recruited) and are grouped according to genotype for the data analysis. The data analysis can then determine if the phenotype values differ significantly among the various genotype groups (typically two or three groups are compared).

The other typical design scenario is to recruit subjects with specific phenotype values and identify possible genotype and allele frequencies underlying those different phenotype values. This is known as

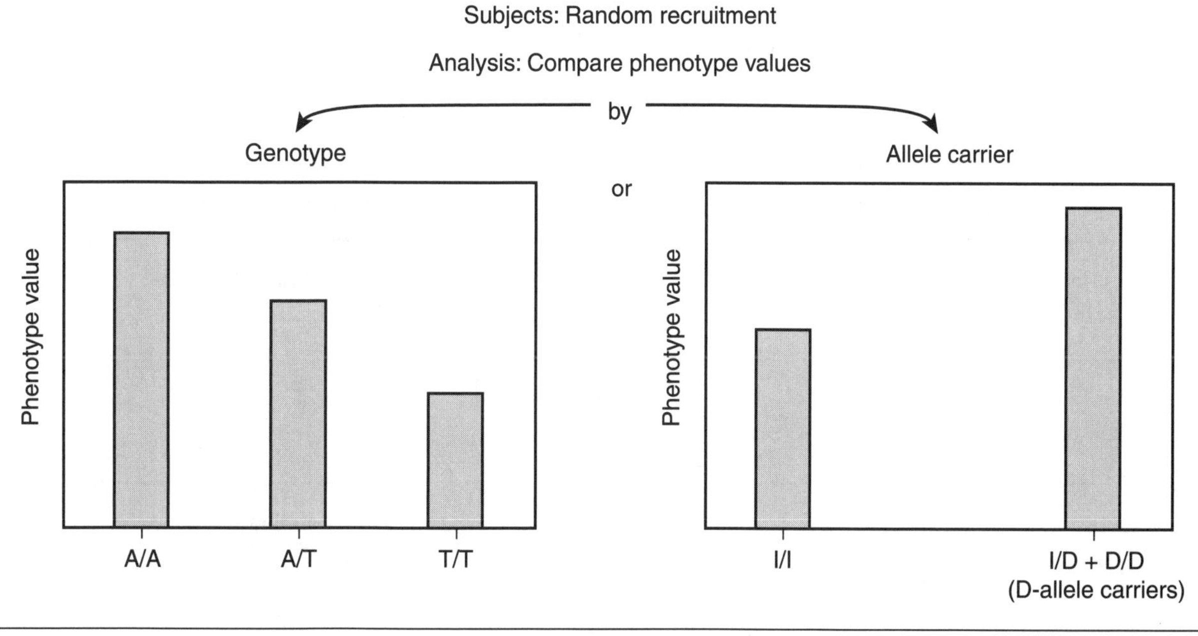

Figure 9.1 The typical association study designs used to address genetic association hypotheses. Subjects are not selected for a particular phenotype; rather the phenotype value is compared among genotype or allele carrier groups to identify phenotype differences associated with different genotypes.

a **case–control design** and is seen most often in studies of genetic disease. Traditionally, a *case subject* is thought of as an individual who carries a specific disease phenotype (e.g., muscular dystrophy), while a *control subject* is a similar individual (e.g., perhaps age and sex matched) who does not carry the disease (i.e., a healthy subject). More broadly, we can think of cases and controls as simply having different levels of the phenotype of interest. For example, we could recruit otherwise similar individuals with high versus low $\dot{V}O_2max$, or high versus low blood pressure, and so on. In all of these examples, the recruitment is not typically random, focusing instead on differences in the phenotype itself. In the case–control design, the hypothesis is that genetic factors have contributed to the presence of a high or low trait value, and thus different allele frequencies would be expected in the case versus control samples. An example is shown in figure 9.2, where allele frequencies are determined for both a case and a control population and the frequencies are compared in order to identify any significant differences. For a G/T polymorphism, if the G allele contributes to a disease phenotype or higher values for a particular trait measurement, we would expect more G/G individuals in the case group (i.e., higher G-allele frequency) than in the control group, where we would expect a higher frequency of the T allele.

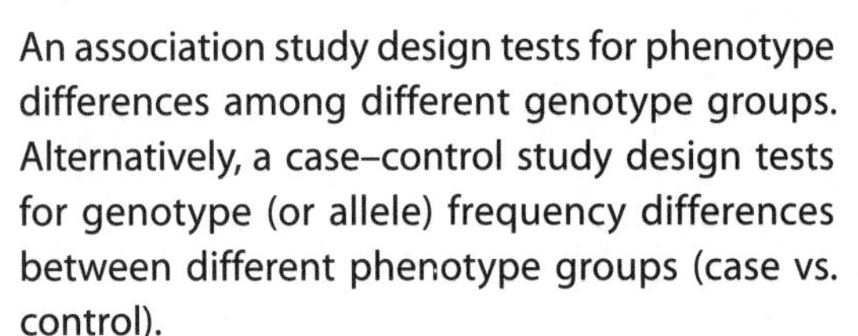

KEY POINT

An association study design tests for phenotype differences among different genotype groups. Alternatively, a case–control study design tests for genotype (or allele) frequency differences between different phenotype groups (case vs. control).

A variation on this theme is shown in figure 9.3, where cases and controls are assigned within a randomized sampling of subjects. In other words, random subject recruitment occurs for all phenotype levels, as in a typical genetic association study design, but a case–control type of design is developed from the extremes of the phenotype distribution within the recruited subjects. The cases and controls are assigned after recruitment and measurement of the phenotype in all subjects. This is typical when extremes of a nondisease phenotype are examined, as was described previously for a case–control analysis of $\dot{V}O_2max$ or blood pressure. An alternate name for this type of design is *high trait/low trait*.

Other study designs are obviously possible, but these are the ones you'll most commonly see in the research literature in human genetics. I chose not to focus on family-based designs in this section, as these are covered briefly in other areas of the text. Let's now move to discussing the actual recruitment of subjects into a genetic association study.

Subjects: Recruit for phenotype (affected and matched unaffected)

Analysis: Compare allele frequencies

Genotypes of cases

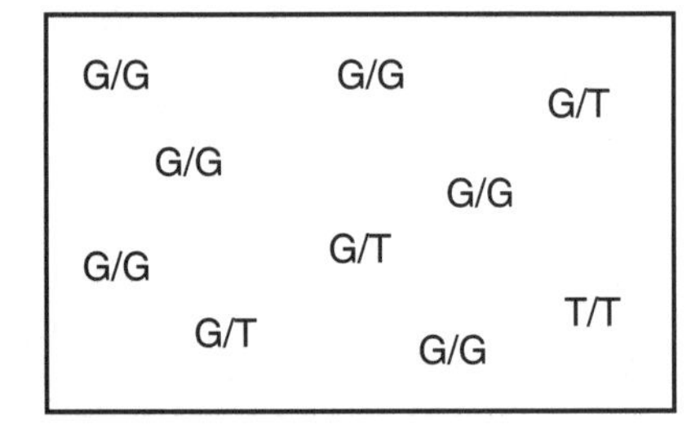

Allele frequencies:
G: 0.75
T: 0.25

vs.

Genotypes of controls

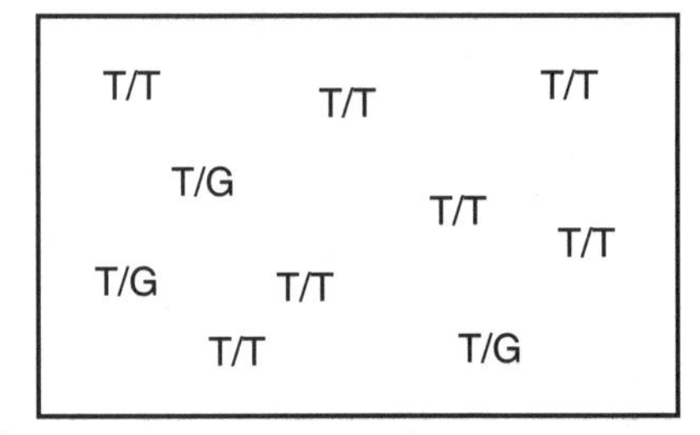

Allele frequencies:
G: 0.15
T: 0.85

Figure 9.2 The typical case–control study design, in which subjects are recruited for the presence or absence of a particular phenotype (e.g., disease) and the allele frequencies are compared between cases and controls to identify genetic association. In this example, the genotypes of 10 cases and 10 control subjects are shown in each box, with the allele frequency calculated for both.

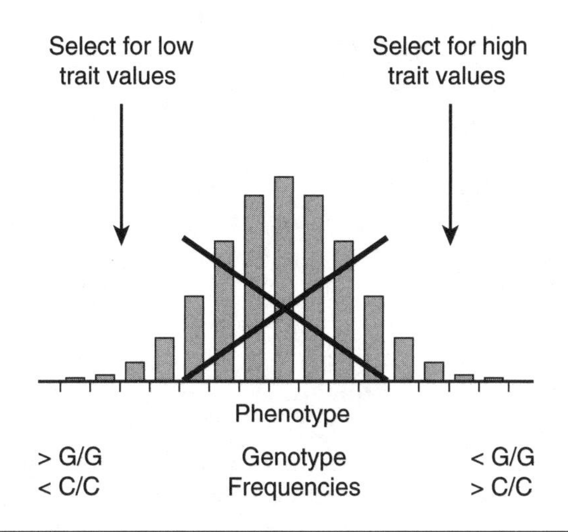

Figure 9.3 If subjects are recruited regardless of phenotype (e.g., a nondisease quantitative trait), the subjects with extreme measures of the phenotype can be used to generate a case–control type of study design.

SUBJECT RECRUITMENT

With the introduction of a genetic variable into a study, the biggest impact on the study is seen in how the subjects are recruited or grouped (or both) for statistical analysis. If we design a case–control analysis, the focus is on defining phenotype differences and recruiting specifically to meet those phenotype definitions. For genetic association study designs in which subjects are grouped by genotype, the primary impact on subject recruitment comes from the genotype and allele frequencies for the polymorphism(s) chosen for investigation.

Anticipate Sample Sizes

If we focus simply on single-nucleotide polymorphisms (SNPs) and insertion/deletion polymorphisms (repeat polymorphisms add additional complexity), rarely do we find a polymorphism in which the frequency of each of the two alleles is 50%. Very often, the rare allele frequency is somewhere in the range of 15% to 45%, which means that the frequencies for the three genotype groups will not be equal. To calculate expected genotype frequencies (genotypes: AA, AB, and BB) from the allele frequencies (A and B alleles), we use the following calculation: %AA = (A-allele frequency)^2; %BB = (B-allele frequency)^2; %AB = 2 · (A-allele frequency) · (B-allele frequency). These calculations are covered more completely in appendix B, "Evolution and Hardy-Weinberg Equilibrium." For example, if the allele frequencies for an A/G SNP are 75% and 25%, the expected genotype frequencies will be 56% (0.75^2), 38% ($2 \cdot 0.75 \cdot 0.25$), and 6% (0.25^2), for the A/A, A/G, and G/G genotype groups, respectively. Thus, if we were to randomly recruit 100 subjects into a study to investigate the association of these genotypes with a phenotype of interest, we would expect 56, 38, and 6 individuals within each of our genotype groups, respectively. Are 6 subjects with the G/G genotype enough for valid statistical analysis? Although many of the statistical procedures used in any data analysis are able to handle slight deviations in sample size across compared groups (i.e., *unbalanced* sample sizes), very unbalanced samples can cause problems. While the details of statistical analysis are beyond the scope of this text, I note here that it is important to anticipate sample sizes before recruiting subjects and to understand any implications for the statistical analysis that will be performed on those groups.

Approaches to Recruitment

Once a polymorphism is identified and chosen for a study, there are several typical scenarios for subject recruitment that can result, depending on the expected allele and genotype frequencies (as determined in the genome databases or from previous research). In the first scenario, the expected allele and genotype frequencies may be such that a reasonably balanced sample can be recruited using standard, random recruitment methods (e.g., random sampling from a population). This is the preferred scenario, as it affects study design and data analysis the least when compared to study designs that don't include genetic factors. An example is the insertion/deletion (I/D) polymorphism in the angiotensin-converting enzyme (ACE) gene, in which the I-allele frequency is typically reported to be between 45% and 48%; here the researcher will achieve a reasonably balanced sample simply from a random recruitment strategy.

In a second scenario, the allele frequencies are such that unbalanced sample sizes are expected and the statistical analysis may be problematic because the genotype group sizes will be too different. An example could be the A/G SNP described earlier with a rare allele frequency of 25%. In this case, two options exist: (1) during recruitment, screen subjects by genotype to ensure balanced group sizes, or (2) perform a random recruitment, but group subjects by allele rather than genotype.

KEY POINT

Subject or participant recruitment can be complicated when the polymorphism of interest has a low rare allele frequency. In these cases, prescreening subjects for genotype is often necessary to result in balanced group sizes.

The first option (screening for genotype) is more labor-intensive, in that prospective subjects must provide a DNA sample at the beginning of a study, which must then be tested in the laboratory so that genotype can be determined for each subject. Then, from all of the subjects initially recruited, all of the rare allele homozygotes are typically chosen to complete the study, while some smaller fraction of heterozygotes and common homozygotes are chosen, resulting in balanced sample sizes among the genotype groups that complete the study. If we use the previous example, in which allele frequencies for an A/G SNP are 75% and 25%, the expected genotype frequencies will be 56%, 38%, and 6% for the A/A, A/G, and G/G genotype groups, respectively. If we want a balanced sample size, such that ~33% of subjects are in each of the three genotype groups, and we want 100 subjects in each group, then we will need to screen a total of 1667 individuals for their A/G SNP genotype in order to ensure 100 G/G subjects (only 6 of whom are found in every 100 people!). From these recruited subjects, we will select the 100 G/G subjects for inclusion in the study and then randomly select 100 subjects from each of the A/G and A/A genotype groups to complete the study design. Obviously, the time and effort to produce such a balanced group size are considerable, so researchers do not make this decision lightly! Obvious statistical challenges, or perceived benefits in terms of data interpretation, must be evident in order to justify taking on the additional costs of genotype prescreening during the recruitment phase.

The second option when the rare allele frequency is low is to group subjects according to allele rather than according to genotype. In the case of SNPs and insertion/deletions, the most common result is that instead of three genotype groups, researchers end up with two groups for data analysis purposes. Using the example we have been discussing, if we expect that genotype frequencies will be 56%, 38%, and 6% for the A/A, A/G, and G/G genotype groups, respectively, then from 300 recruited subjects we will expect sample sizes of 168, 114, and 18, respectively. If the size of the G/G genotype group is considered too small for proper statistical analysis, then many times researchers will combine the A/G and G/G groups (i.e., 114 A/G + 18 G/G) because both of those groups contain the rare G allele. In other words, rare allele carriers (A/G and G/G subjects) are grouped together and compared against the common-allele homozygotes (A/A subjects). The result would be two groups with sample sizes of 168 (A/A) and 132 (A/G + G/G), which is much more balanced for statistical purposes.

The benefit of this course of action is that the incredible cost and time associated with genotype prescreening are eliminated; but the limitation is that any differences between the A/G and G/G groups cannot be determined, decreasing the ability to interpret any positive genetic associations. For example, if the phenotype value for the A/G + G/G group is significantly different from that for the A/A group, is the G allele dominant or recessive? This question is harder to answer when the A/G and G/G groups are combined. Moreover, if the G allele is recessive and there are relatively few subjects with the G/G genotype, the statistical analysis could reveal no difference in a comparison of A/A versus A/G + G/G groups. Hidden within such a result could be a very real and interesting recessive trait observed only in the G/G carriers. Alternatively, the recessive trait could result in the A/G + G/G group's being significantly different from the A/A group, leading researchers to conclude that the G allele (by itself) was influential on the trait. In this case, the researchers would have wrongly inferred an influence of the G allele because they were not able to observe the recessive influence of the G/G genotype. Thus, combining allele carrier groupings in a statistical analysis should be approached with caution.

A final scenario is that the rare allele frequency is low enough that a screened recruitment is considered too challenging, and grouping subjects by allele (e.g., AA vs. A/G + G/G) is either statistically problematic or too limiting from an interpretation standpoint. In this case, the polymorphism is removed from the candidate gene list to be replaced with another one that is more amenable to subject recruitment issues.

GEOGRAPHIC ANCESTRY

An additional concern for subject recruitment in genetic association studies is the issue of race and ethnicity within the subjects under study. This is a very complex and controversial issue, discussed further in the Special Focus section in this chapter. The underlying issue is that populations with different geographic ancestries (e.g., African, Asian, Northern

European/Caucasian) have different allele and genotype frequencies for some but not all polymorphisms. Populations from different geographic origins will carry unique proportions of various alleles (though few alleles are specific for any one geographic ancestry). Thus, all individuals, regardless of their self-identified race category, carry some history of their geographic ancestry in their DNA sequence. Think of **geographic ancestry** as a definition of a *very* broad family, such that members of that geographic family share a greater fraction of their DNA sequence variation with one another than they do with members of other geographic families.

Why are these differences important? The issue really comes down to the idea of gene × gene interaction (discussed at the end of chapter 6)—that alleles in one gene interact with alleles in another gene, and the resulting interaction (the combination of alleles) will have specific influences on the trait of interest. Often (because of statistical limitations), researchers can focus on only one or two genes in a single study, so the question of gene × gene interaction cannot be addressed because all possible interacting genes and alleles cannot be included in a single analysis. Instead, researchers are forced to focus on one or two genes while knowing that other genes are going to be important. Because subjects with the same geographic origins share some fraction of DNA sequence variation across the genome, much like a very extended family, it is possible that certain gene × gene interactions will be observed more often in certain geographic populations than in others, because the frequency of various alleles differs in those different populations. Thus, different genetic associations may be identified in different populations because of these unique combinations of alleles. By studying or analyzing individuals according to their geographic ancestry, researchers may gain some ability to predict unique gene × gene interactions without having to study the specific genes and alleles important to the interaction. Because in any research study subjects will likely come from different geographic origins, accounting for differences in geographic ancestry may be necessary to test for unknown gene × gene interactions. Conversely, when one studies all subjects regardless of geographic ancestry, important gene × gene interactions occurring in one population subgroup may not be evident because of their inclusion with the other population groups in the study.

Here, we should distinguish between two commonly used terms, race and ethnicity, that are often used interchangeably but rather represent distinct categorizations. "Race" is most often used to distinguish populations of people based on physical differences, which are presumably based in slight genetic differences. "Ethnicity" is similarly used for grouping individuals, but the term is focused more on cultural similarities (e.g., behavior, language, religion) than on biological traits. In many cases, the terms race and ethnicity may overlap for a group of people; but for the purposes of defining a population's genetic predisposition to a disease or other phenotype, race is the most appropriate term (see complete discussion in the Special Focus section).

Unfortunately, the most efficient means of determining geographic ancestry is an imperfect one: the use of self-identified race categories. In other words, by asking research subjects their race, we have a rough but imperfect means of gaining insight into their geographic ancestry. But using race as a variable, which by itself is not a biological variable but rather a social construct, is highly controversial because of the potential for race discrimination (discussed more fully in the Special Focus section). Thus, researchers are in the challenging place of knowing that potentially important biological information is imperfectly correlated with a socially derived grouping variable. I wish I could tell you that there was a consensus in the field about whether or not to use race in such a way, but there is no such consensus; researchers fall on both sides of the issue. The bottom line is that you should consider this when recruiting subjects and performing analyses, because regardless of whether you personally agree or disagree with the use of race as a grouping variable, your colleagues (or those reviewing your papers and grants) may disagree with you. Thus, consideration must be given to whether information about race should be collected as a grouping variable and, once collected, how it should be dealt with in the analysis.

Two terms common in genetics research are related to this issue of geographic ancestry. The first is **population stratification,** which refers to the idea that within any group of people (especially a diverse group of people), each individual will have some unique DNA sequence that is correlated with his or her geographic ancestry. Population stratification is a term used to describe the extent of such differences among individuals in a group, according to their geographic ancestries, with very diverse groups described as highly stratified.

The reason that race is an imperfect means of identifying geographic ancestry is that mating occurs across different populations, thus DNA sequences from different ancestral regions are mixed in any particular individual—more so in some, less so in others. In other words, there are no strict "boundaries" that distinguish one race or geographic ancestry group from another. This second concept is known as **admixture,** which is the extent to which DNA from different geographic origins is combined in any individual or subpopulation. Thus, if researchers think it is important to account for geographic ancestry in the analysis of a particular phenotype, then subject recruitment will have to fulfill this need by including geographic ancestry (typically by using race) as a categorical variable. Highly stratified or highly admixed populations will be far more challenging to study from the perspective of geographic ancestry, because each individual subject will be less easily categorized as coming from any single geographic origin.

SPECIAL FOCUS
Race in Genetics

Should race be included as a classifying variable in genetics research? This is a controversial question, with scientists falling on both sides of the debate. The rationale for even considering race as a variable stems from the fact that various diseases and disease risk factors occur more frequently in some races compared to others. For example, sickle cell anemia is a disease common in individuals of African descent, but rare in East Asian and Northern European/Caucasian populations. Prostate cancer is more prevalent in African American than in Caucasian men. East Asians are more often susceptible to side effects of various drugs than other races because of a tendency for slower drug metabolism. Many other examples of such health disparities exist among various race and ethnic groups. These health disparities are the focus of intense research, aimed at identifying the underlying disease mechanisms so that the disparities can be reduced or eliminated altogether. In many cases, differences in environmental factors (socioeconomic status, access to health care, cultural diet preferences, exposure to different pathogens, etc.) can be pointed to as a basis for many health disparities, or at least an explanation of some part of a disparity among populations. In some instances, though, disease susceptibility rates continue to differ among race groups even after such environmental factors are accounted for, indicating a potential biological mechanism.

The reason for the controversy around the inclusion of race in such work is that race by definition is a social construct, not an inherently biological trait. In other words, while geographic ancestry is correlated with race, there are no true boundaries or physical traits that uniquely define an individual as being of one race or another. And while various environmental and cultural similarities may exist for members of a race group, in no way can it be assumed that all individuals identifying themselves as a certain race will be affected by those common cultural and environmental factors. As discussed in the main text, genetic differences do exist in populations of different geographic ancestry, and many individuals can be grouped within major race categories based on an analysis of their DNA sequence. But while geographic ancestry is associated with minor biological differences among populations, race in itself is an imperfect means of studying those biological differences. This background, brief though it is, can be used for a better understanding of the arguments on both sides of the question of including race in genetics research specifically and health-related research in general.

Those scientists who argue against the inclusion of race as a variable argue that race is an imperfect means of identifying geographic ancestry, too imperfect to be useful for understanding health disparities. Though differences in allele frequencies do exist among populations of different geographic ancestries, the vast majority of genetic variation occurs across all populations, with little variation occurring in one but not another group. In other words, there are few polymorphisms that are specific to any particular race group—limiting the idea that extensive genetic differences exist among different race groups. Moreover, phenotype variation among humans is continuous: There are no conclusive boundaries between one race and another, and thus there is no way (e.g., skin color) to consistently classify race for any individual. Trait variability underlies the concept of race. Most often, researchers rely on self-report as a means of racial classification, with individuals being asked to choose among a small number of race and

(continued)

SPECIAL FOCUS

(continued)

ethnic categories, and such classification may not be entirely accurate. Finally, it has been argued that with the use of race as a variable in health-related research, the idea that races are inherently different and therefore potentially unequal from a biological standpoint is perpetuated, with the risk that discriminatory social practices will be continued and thereby justified by small-minded individuals.

On the flip side of the argument, some scientists contend that while race is imperfect as a measure for geographic ancestry, accounting for geographic ancestry is important when one is examining health disparities, and race represents a means of accounting for these ancestral differences. Moreover, genetic differences do exist among the various race groups, such that individual DNA samples can be reasonably categorized into one of a few race groups with decent accuracy. In other words, while most genetic variation occurs in all race groups, the frequency of some alleles differs dramatically and consistently across different race groups—arguing that including race as a categorical variable will help account for some of these consistent genetic differences. Finally, it can be argued that even after known environmental factors are accounted for, disease risk does differ among race groups, indicating that biological factors underlie some health disparities.

Ultimately, both sides agree that the goal of moving forward with research in health disparities is to identify the specific environmental and biological factors (e.g., genetic variants) that explain differences in disease risk observed among individuals, regardless of their self-identified race. In other words, the goal is to study disease risk at the individual level, rather than at the group level. While various technologies are being developed to move researchers closer to this goal (e.g., SNP chips for broad assessments of genetic variation), these technologies are not adequate at present (e.g., too costly, imperfect). Thus, many researchers continue to use race as a means of grossly accounting for geographic ancestry in their work, knowing the inherent limitations. As we will discuss in chapter 12, we all eagerly await the time when individual genetic information can be studied in place of race and ethnic categories, which is the idea behind personalized medicine.

PROSPECTIVE VERSUS RETROSPECTIVE

Very often, researchers come to the idea of including genetics in a study after the study has already been started or finished. When genetic variables are added to an existing study, the study is considered *retrospective,* whereas studies designed from the start to investigate specific genetic factors are known as *prospective.* Retrospective studies are possible because of the unchanging nature of DNA sequence: The various alleles you carry today are the same that you were born with and the same that you carried while completing a research study a few years ago. If researchers are able to obtain a DNA sample (from a blood or other tissue sample) as a part of a research study or sometime after a study's completion, regardless of the original intent of that study, the opportunity exists for a retrospective genetic analysis. I'm ignoring at this point the issue of subject consent for performing such an analysis, but that will be covered in chapter 13.

A primary concern with performing retrospective studies is that there is no control over subject recruitment for the issues of allele and genotype frequencies. If you are interested in studying a polymorphism with a low rare allele frequency, you have limited options in a retrospective study. Prescreening of genotype in order to produce balanced sample sizes is no longer an option, so it is important to anticipate genotype group sizes before performing any laboratory work. If the group sizes are going to be severely unbalanced, and analysis by allele grouping (e.g., AA vs. A/G + G/G) isn't justified, it will be necessary to choose a new polymorphism with a larger rare allele frequency. Moreover, recruiting additional subjects is often problematic, so if sample sizes are too small for proper statistical power, a retrospective analysis may not be recommended.

A secondary concern for retrospective studies is that recruitment may not have accounted for geographic ancestry, resulting in samples with considerable population stratification. If accounting for geographic ancestry through race or other nongenetic means is important for your phenotype of interest, then you'll want to make sure such variables exist in the study database before committing to genetic analysis in a retrospective fashion.

Finally, retrospective studies do not have the benefit of careful consideration of phenotype measurement from a genetic variation standpoint. As discussed earlier, the key consideration in a genetic study is to account for as many important environmental factors as possible, thus allowing the greatest opportunity for defining phenotype differences due to genetic factors. Were all of these important environmental factors measured in the original study? Were the environmental factors measured in an accurate way across all of the individuals of the study for inclusion within statistical models? Was the phenotype itself measured as you would measure it in a prospective study, or was the measurement not optimal (e.g., was it an older technique with greater error, or was it of a broader phenotype rather than a more specific intermediate phenotype)? Very often, one can deal with these issues when adding genetic analysis to an otherwise strong research study, but it is advisable to address these questions before resources are committed to the laboratory work needed to get the genotype data.

STATISTICAL APPROACHES

After one has committed to a specific research design, recruited and tested subjects, obtained genotype data, and so on, the next phase of a research study is the statistical analysis. As this is an introductory text, and certainly not a textbook in statistical genetics, my goal here is to present the primary statistical approaches used in typical genetic association analyses, and even then not to discuss them in statistical terms. In other words, the reader should consider this section as an overview of the analysis of genetic data in an association study, not as instructions for performing those analyses.

Correct Labeling

Once the data have been collected, a critical first step is to make sure the genotype data are correctly labeled. This may seem obvious, but because there are no strict standards for genotype labeling, mixing up the alleles for a specific polymorphism is unfortunately not too difficult. Moreover, many genotyping methods provide only pictures from which genotypes are identified in the laboratory, and most fluorescent technologies require that the technician tell the computer software which allele to assign to a particular fluorescent tag. In other words, human error can and does occur at the stage of assigning specific genotypes to each subject in the database. Be sure that the alleles or genotypes that you've assigned are accurate. Go back to your sequence-verified control samples, and also confirm which DNA strand was sequenced (each strand will have different, complementary alleles). Are the alleles that you've assigned for a particular polymorphism the same as those presented for that polymorphism in the genome databases? Do your allele frequencies match those of previous research or the polymorphism databases for your population? These simple checks are critical to preventing a mistake that is easy to make yet embarrassing at the time of manuscript or grant submission.

Another initial step often taken is the calculation of Hardy-Weinberg equilibrium for the polymorphisms investigated in the study, as discussed in appendix B, "Evolution and Hardy-Weinberg Equilibrium." While this is not often viewed as a critical step in a genetic association study, many researchers insist on this calculation despite its generally limited usefulness. Do the genotype frequencies obtained within your research sample match those expected under the conditions set forth by Hardy-Weinberg equilibrium? There is considerable controversy about whether this is an appropriate test to be performed in all genetic association studies, because in many cases it is used inappropriately to test for genotyping accuracy. Nevertheless, it is frequently calculated and reported for genetic association studies. See appendix B for additional details.

Groupings and Statistical Tests

Once the genotype and other data have been checked thoroughly and normalized as might be appropriate, subjects are grouped according to the criteria set out in the study design. Cases and controls are grouped for comparison of allele and genotype frequencies, or subjects are grouped by genotype or allele for comparison of phenotype values among the groups. In other words, there are no special analyses that must be performed in a genetic association study simply because genetic data now form the basis of the group comparisons. In fact, traditional statistical tools, such as T-tests, chi-square analysis, analysis of variance (ANOVA), analysis of covariance (ANCOVA), and regression modeling, are used similarly to the way they are used in any other research design. In many typical nongenetic studies, subjects are grouped according to some intervention variable and compared using standard statistical tools. For example, subjects are grouped according to sex, or medication status, or exercise training history, and so on. Here, we are doing the same sort of grouping, but the groups are based on phenotype levels or, more commonly, genotype and allele. And, the results of the statistical analyses, whether different at the $P < 0.05$ level or some other standard, are interpreted for a genetics investigation just as they would be for any other type of statistical analysis.

KEY POINT

Despite the inclusion of genetics in an association study design, the appropriate statistical analyses often follow exactly the same models as in those studies designed without a genetic component.

Recall that a major goal of the study design was to either control for or measure the various environmental factors known to influence the phenotype of interest. If the experiment could not control those factors in the recruitment or design phases of the study (e.g., recruiting only age-matched subjects), then the statistical models used in the analysis, especially ANCOVA and regression, will include these environmental factors as covariates. In other words, measurements of smoking status, total caloric intake and caloric expenditure, age, sex, and so on become statistical covariates in the models, thus statistically accounting for known environmental contributors to the phenotype. Some potential covariates, especially sex and race, may require inclusion as independent variables in the model. Men and women are different, and their physiological differences may present unique enough environmental factors that testing them together is inappropriate. Similarly, if multiple races are included in a study and race/ethnicity, as a rough connection to geographic ancestry, is considered of interest to the investigators, different races may require formal statistical comparison.

After all of the hypotheses have been tested, if positive associations have been identified in the statistical analysis, researchers very often want to estimate the extent to which a particular polymorphism is contributing to the total variance of the trait of interest. In other words, just how much of the trait variation is explained by this polymorphism? A typical way to test this is to perform a regression analysis of the full model (i.e., containing all significant independent factors and covariates) and determine the *r*-value of that model, which is an estimate of how much trait variation is explained by *all* of the variables in the model. By repeating this procedure with a regression model that *excludes* the genotype variable (i.e., the full model minus one variable), the researcher will obtain a new, presumably smaller, *r*-value. The *r*-value is expected to be smaller for the revised model, assuming that genotype is explaining some amount of the trait variation. Simply comparing the *r*-values of the two models will provide an estimate of the extent to which a particular polymorphism is contributing to the trait variation in a given study. Recall from previous discussions that the contribution of a single polymorphism to a complex phenotype will likely be small, perhaps only 1% to 2% of the trait variability. More sophisticated means of performing this type of analysis exist, but this is the general idea behind estimating the contribution of a genetic factor to a phenotype.

INTERPRETATION OF RESULTS: ASSOCIATION VERSUS CAUSE AND EFFECT

While the interpretation of the statistical tests in genetic studies is the same as in any other study, what the final results mean for a genetic study is another matter. You will need to ask certain questions when interpreting the results of genetic studies. Did you find a positive, significant association between genotype and phenotype? Was the finding in the entire cohort, or simply in a subgroup (e.g., women only)? What covariates were also significant contributors to the statistical models? What proportion of trait variability was explained by genotype? Once these questions have been addressed, do the results fit with any specific, directional hypotheses that may have driven the study (as opposed to general, nondirectional hypotheses)?

Once you've thoroughly examined the results of your study, critically examine any previous work in the research literature. Have others reported similar associations? If so, what populations or subgroups did they study? What covariates were included in their models, and were their models similar to those tested in your study? How do your results fit with the existing literature, and is consistency observed among the related studies?

Why spend all this time discussing fairly obvious questions? *Because genetic studies are studies of association, not studies testing cause and effect.* We are not changing the genome sequence of our subjects and observing the results on a particular phenotype, which would constitute a cause-and-effect type of study (and is typical of many animal genetic studies). Rather, we are making observations about correlations among genotypes and alleles with phenotype measurements. We cannot claim cause and effect in human genetic association studies! Moreover, in many cases the allele of interest may not have been tested for a functional influence on the gene or protein within which the allele lies. In other words, we often don't know if the associated allele is biologically meaningful. As we discussed in detail in chapter 7, an associated allele might be only a "marker" for a true functional allele in a nearby polymorphism (this is the concept of *linkage disequilibrium*). On the other hand, animal models, in which direct genetic manipulations are possible, do provide an opportunity for cause-and-effect studies, though sometimes there are concerns about differences between the physiology of the animal model and human physiology.

The bottom line is that one must take extreme care when interpreting and discussing the results of a genetic association study. Unless there is strong evidence to suggest that the allele is functional (e.g., verified nonsense polymorphism, predicted missense SNP, promoter polymorphism previously shown to affect gene transcription), claims of genetic association must be made with caution. For functional alleles, speculation about cause and effect may be warranted, depending on the strength of the evidence showing influence of the allele on gene or protein function. For alleles of unknown function, the general strategy is to claim that the basis for the association is unclear, the function of the allele in question is unknown, and the allele may simply be acting as a marker for a nearby, functional allele. Ultimately, the replication of a genetic association, either by the same research group in a new, independent sample of research subjects, or by another group in an independent sample, is a critical component to moving from nonspecific genetic associations to a more complete understanding of the functional role of a particular allele on a phenotype.

SUMMARY

We've covered considerable ground in this chapter, from the basics of human genetic research design all the way to the general interpretation of results. Subject recruitment in genetic association studies is often problematic, because subjects are frequently grouped by genotype, and rare allele frequencies may prevent easily balanced sample sizes. Moreover, issues of population stratification must be considered for some phenotypes, though the inclusion of race as a categorical variable is a topic of considerable debate. Whether to study cases and controls or use a randomized sample is another consideration. Once the study is designed and completed, statistical procedures are fairly straightforward, but the interpretation of genetic association studies must be performed with caution: Association is *not* cause and effect.

KEY TERMS

admixture
association study design
case-control design
geographic ancestry
population stratification

REVIEW QUESTIONS

1. How do the two major types of study design—the genetic association and the case-control design—differ? What is the focus of subject recruitment for each type of design? How can a case-control type of study be developed out of a typical genetic association study?
2. How does rare allele frequency affect subject recruitment procedures for a genetic association study? Calculate the expected genotype frequencies for a SNP with a rare allele frequency of 20%.

3. Why is race sometimes considered in health-related research, despite the fact that it is a social rather than a biological variable? What are some arguments against the inclusion of race in medical research?
4. In what ways is it challenging to add genetic analysis to a study that has already been completed?
5. Can cause-and-effect conclusions typically be generated from human genetic studies? Why?

CHAPTER

10

BASIC LABORATORY METHODS IN GENETICS

We end part II of the book with an overview of how genetic variation is measured in research or clinical settings. As has been emphasized throughout the book, this text is meant to provide an overview of the field of genetics rather than a detailed examination of all aspects of the field. Similarly, this chapter should be regarded as an overview of some of the major techniques used to study genetic variation, not as a comprehensive guide to the many laboratory methods used in even basic genetic analyses. The intent is to provide readers with a basic understanding of the methods used in human genetics work, as are often discussed in the methods sections of genetic association research papers.

DNA FROM BLOOD OR CHEEK CELLS

Ultimately, any genetic investigation (with the exception of family-based correlation studies) requires DNA. Fortunately for researchers and clinicians, DNA is easy to obtain, as it is present in nearly every body cell.

The most common source of DNA for a genetics study is a blood sample. Even a small drop of blood contains enough DNA for genetic analysis, though the most common practice is to obtain a blood sample from an arm vein. As shown in figure 10.1, when a whole-blood sample is centrifuged, or spun at a high speed, the three major components of blood separate into three distinct sections. The top layer (the lightest of the layers when spun at high speed) is plasma, the liquid component of blood. The middle layer, often called the **buffy coat,** is a thin layer of *white blood cells***.** This is the major DNA source in blood. The bottom layer is composed of the *red blood cells* (RBCs). In the process of formation of RBCs, the cell nucleus is discarded such that the RBC is simply a hemoglobin-carrying (and thus, oxygen carrying) cell with no DNA component; thus RBCs cannot be used for obtaining DNA.

Once a blood sample is obtained, the sample is centrifuged and the buffy coat of the sample is removed. While small amounts of DNA are also found in the plasma, the greatest quantity of DNA is found in the white blood cells of the buffy coat layer. In order for genetic studies to be performed, the DNA of these cells must be separated or extracted from the other cell components. To this end, the buffy coat's cells are removed from the sample and placed in a separate tube. The sample is cleaned of any residual RBCs and plasma, leaving a clean complement of white blood cells. The membranes of these cells are then broken up (or "lysed"), such that both the outer cell membrane and inner nuclear membrane are dissolved and the DNA material is now suspended in the tube. In the final steps of the **DNA extraction** process, the DNA is cleaned of all of its associated proteins (such as those used for packaging DNA into condensed chromosomes) and isolated, resulting in pure DNA molecules. The pure DNA can then be studied using a variety of procedures, as will be described later. The DNA extraction process is a fairly

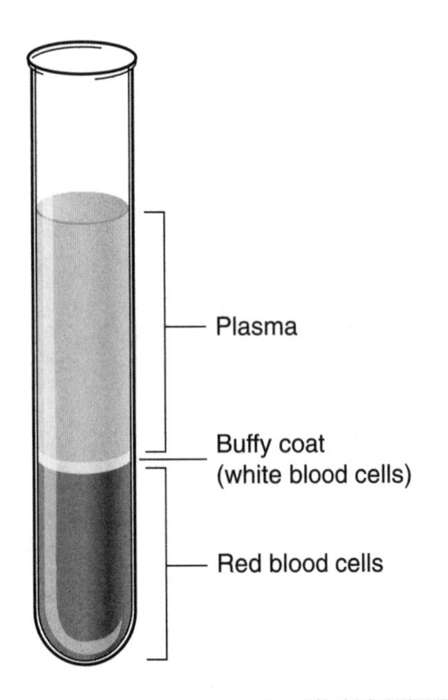

Figure 10.1 The major components of whole blood, shown after a typical centrifugation of a blood collection tube. The bulk of the DNA is contained in the buffy coat, or white blood cell layer. Red blood cells do not contain DNA.

straightforward procedure at a lab bench, requiring typical lab equipment and chemicals. The extraction of DNA takes about 2 hours.

Fortunately, blood samples are not the only sources of DNA. Another common source of DNA material is the cells of the cheek inside the mouth, known as **buccal cells.** A cotton swab vigorously rubbed against the inside of the cheek, or a mouthwash solution vigorously swished inside a clean mouth, can remove enough buccal cells that DNA can be collected in sufficient quantities for genetic studies. This option is often used when research subjects are far from the research setting and a DNA sample must be mailed, or it can be used in children when a blood sample might be difficult to obtain. The whole-blood DNA sample is preferred, however, as the quantity and quality of DNA available in a blood sample are often superior to those provided by buccal cells.

KEY POINT

For most research purposes, DNA is extracted from the white blood cells within a whole-blood sample or from the cells of a buccal or cheek swab sample.

Once extracted, DNA is a remarkably stable molecule. DNA can be stored for years at 4° C, or typical refrigerator temperatures. Moreover, the standard techniques used in a genetics laboratory allow the DNA from a single tube of blood to be used in hundreds of genetic tests. These two attributes, long storage life and great quantity, allow genetic analyses to be performed long after the DNA sample has been obtained from a person. This has the positive implication that the person under study need not provide a new sample for every test being performed, but rather can provide a single sample that can be stored for future use. The negative implications are ethical and will be explored in chapter 13.

Why would it benefit researchers to have a DNA sample that could be studied many times far into the future? The answer lies in the number of genetic variants found across the 20,000+ genes in the genome. Each of the millions of polymorphisms could potentially be of interest to scientists trying to understand the influence of genetic variation on various traits. Even for a single research team focused on a single phenotype, there are often tens to hundreds of genes of potential interest, with many polymorphisms within each of those genes. Thus, the number of tests that could be performed on a single DNA sample could be quite large. Therefore, conserving a DNA sample from a research subject for future investigation is crucial for researchers, and laboratory techniques have been developed to study DNA in a very efficient and conservative manner.

POLYMERASE CHAIN REACTION (PCR)

If we could point to a single technique in the toolbox of genetics laboratories as critical for the conservative use of DNA, that technique would be PCR. The **polymerase chain reaction (PCR)** is a means of copying small quantities of DNA into vastly larger quantities of DNA. In effect, the technique is a copy machine for DNA. The procedure is not designed to copy the entire sequence of DNA, but rather to copy only small, targeted regions of DNA sequence. For example, a clinician interested in learning a person's genotype at a polymorphism in exon 2 of a gene could target the PCR technique for that small stretch of DNA surrounding the polymorphism. The PCR procedure would provide millions of copies of that small sequence of DNA, producing ample quantities of material for the clinician to study using other techniques. Hypothetically, a single DNA molecule could be copied into millions of molecules using PCR, though in standard practice several hundred DNA molecules act as the starting material for the PCR assay. Because of this profound capacity to produce large quantities of DNA from very small starting quantities, PCR is arguably the most widely used technique in today's molecular biology laboratories. Polymerase chain reaction forms the basis for the efficient and conservative use of DNA in the laboratory.

As implied in the name of the technique, PCR relies on the **DNA polymerase** enzyme to make copies of the DNA molecule (the DNA polymerase enzyme replicates DNA in the body cells). A small amount of DNA from the stored DNA sample (even as small as tens of nanograms of DNA) is placed in a small tube. To that sample, DNA polymerase enzyme, DNA nucleotides (A, T, G, and C, needed to make the new DNA copies), PCR "primers," and other supporting chemicals are added. This complete chemical soup provides all of the components necessary for the polymerase enzyme to copy the targeted region of DNA sequence. As also implied in the name of the technique, the polymerase does not copy the starting DNA one time, but makes copies over numerous "cycles" in what is known as the "chain reaction" of PCR. As we'll see in a moment, these cycles are initiated and repeated through changes in the temperature of the tube.

Before we examine the details of the PCR reaction, the **DNA primers** must be described. When a specific DNA region has been targeted for analysis (e.g., a polymorphism), the laboratory technician knows the surrounding sequence of that region, usually from the database sequence produced by the Human Genome Project. The DNA polymerase enzyme in PCR requires a starting signal for knowing where to make copies of DNA, and this is provided by a small sequence of 18 to 25 nucleotides known as a *primer*. For each region of interest, two PCR primers are designed as complementary DNA strands that will bind to the DNA sequence a couple of hundred bases on either side of the DNA region of interest. For example, if we want to copy the region containing a polymorphism, we might design a primer that will bind at 100 nucleotides upstream of the polymorphism and another primer that will bind at 100 nucleotides downstream of the polymorphism, thus providing a final copy length of 200 nucleotides. Larger regions can also be copied, and the primers would be designed to be specific to the region of interest and the length of DNA sequence required for the subsequent laboratory methods.

The PCR primers are each designed to bind to one of the two strands of the double-stranded DNA molecule. We accomplish this by imagining each strand as a separate, single-stranded molecule to which we will bind one of our PCR primers. As shown in figure 10.2, we design a complementary PCR "forward" primer in the upstream position by assigning to the primer the bases complementary to the DNA strand where

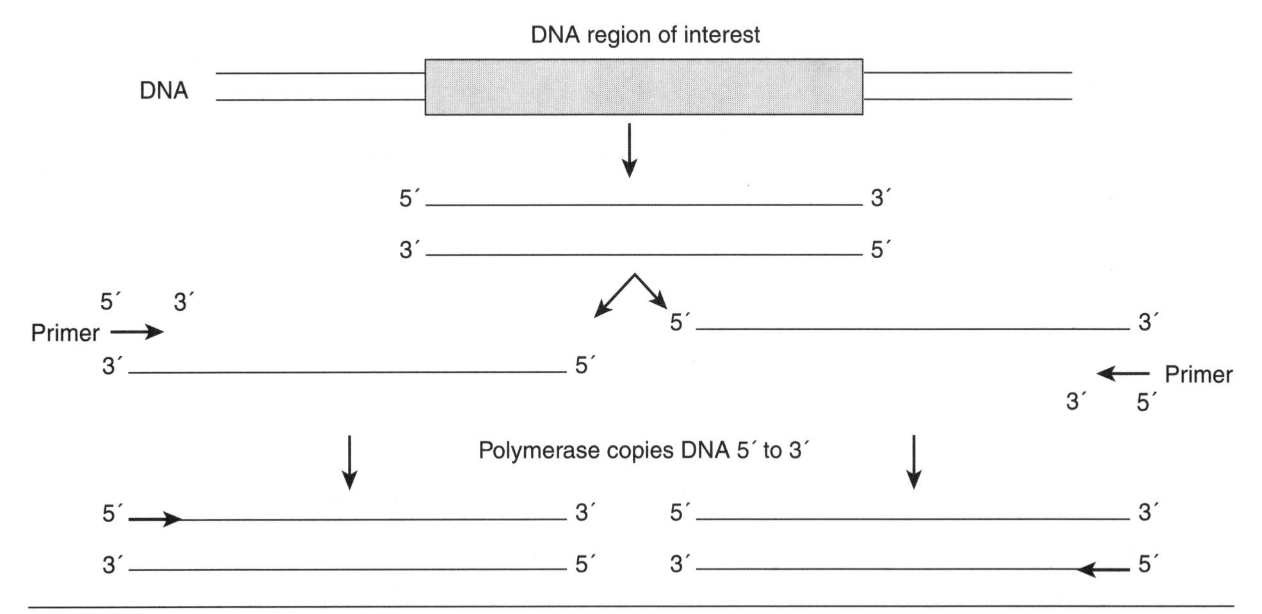

Figure 10.2 An example of how PCR primers can be used to generate copies of a DNA region of interest. When DNA is denatured, or "melted" into single-stranded molecules, the primers (horizontal arrows) can bind to their target sequence, providing a double-stranded starting signal for the DNA polymerase enzyme.

we want the primer to bind. In this way, when the DNA is in single-stranded form, the PCR primer will bind to the DNA, forming a small region of double-stranded DNA. This small region of double-stranded DNA on an otherwise single-stranded DNA sequence serves as the starting signal for the DNA polymerase enzyme. The primer design process is identical for the "reverse" primer on the opposite DNA strand. Thus, two primers are designed, one for each DNA strand on either side of the region of interest. In general, PCR assays are designed to generate sequences of 150 to 700 bases in length for genotyping purposes.

> **KEY POINT**
>
> Polymerase chain reaction (PCR) is a very common laboratory technique used to make many copies of a specific DNA sequence of as few as 100 base pairs in length to as many as several thousand base pairs.

With a rough idea of the PCR primer in place, let's now examine the basics of the PCR process as it happens in the tube, shown graphically in figure 10.3. Once the components of the PCR chemical soup have been put into the tube with the DNA sample we wish to copy, the critical component left for PCR is temperature. Increasing and decreasing the temperature of the tube can alternate the starting DNA material between its double-stranded and single-stranded states. Recall from the design of the PCR primers that the target DNA must be single stranded in order for the primers to bind. Thus the first step of PCR is the **denaturation** step, which is a heating of the PCR soup that results in the separation of double-stranded DNA into single-stranded DNA. The denaturation step is typically about 1 minute at 96° C and brings all DNA in the tube into a single-stranded state. The second step is called the **annealing** step, in which the tube is cooled to between 45° and 65° C (the exact temperature depends on many factors). At this cooler temperature, the small PCR primers bind to the single-

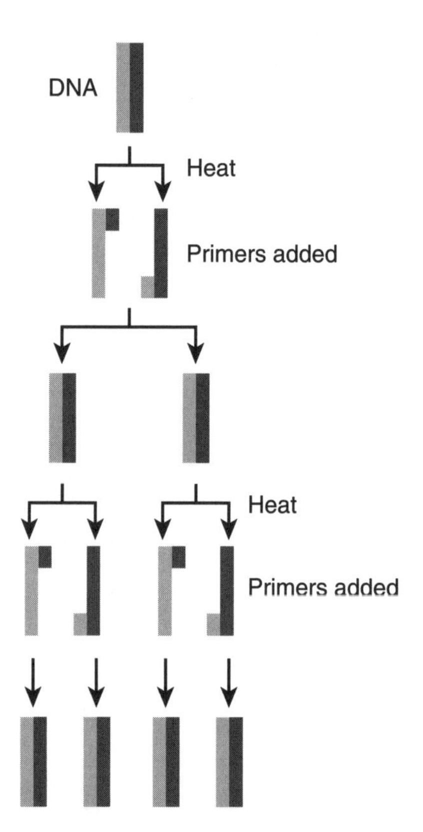

Figure 10.3 An example of the PCR assay, whereby a specific region of DNA sequence can be copied by the use of two specific complementary PCR primers and the DNA polymerase enzyme. Each cycle of denaturation, primer binding (annealing), and polymerase extension results in the formation of new copies of the gene region. Thus, the number of copies increases exponentially until the nucleotides and primers are exhausted in the reaction.

stranded DNA at the region of interest, thus forming small regions of double-stranded DNA. In the third step of PCR, known as **extension** or **elongation,** the DNA polymerase enzyme binds to these primer-bound double-stranded regions of DNA and "fills in" the single-stranded region downstream of the primer at about 72 ° C. The end product is two double-stranded lengths of DNA produced from one double-stranded length that formed the starting DNA material. These PCR cycles are then repeated 30 to 40 times, not only making more copies of the starting DNA, but making copies of the copies, with the end result being an exponential increase in the number of copies in the tube. The final result is often called the *PCR product* to distinguish it from the starting DNA sample.

The description of PCR just presented (and shown in figure 10.3) centered on a single double-stranded DNA molecule as the starting material. In reality, though, there will be several starting DNA molecules in the tube. Making copies of these starting molecules, and then making copies of the copied material in subsequent PCR cycles, can hypothetically produce billions of copies of the DNA region of interest! In reality, the PCR chemical soup contains only so many primers and nucleotides needed to build these copies; so after several million copies are made, the chemical soup is depleted and additional PCR cycles will not yield additional copies. Nonetheless, the 30 to 40 cycles of a typical PCR reaction make enough material for the various "post-PCR" procedures, some of which are described farther on.

Note also that the polymerase enzyme does not stop at a particular region of DNA sequence; rather, it continues to move and copy until either the extension phase ends (i.e., temperature increases as the next denaturation step begins) or the DNA strand ends (as is the case with most PCR strands after the first few cycles). Thus, in early phases, the PCR soup will contain many longer strands of DNA copy than desired, but as more cycles of PCR occur, the polymerase is more and more likely to copy the small PCR fragments (which have multiplied dramatically) rather than the starting DNA material (likely genomic DNA). The end result is an extremely large amount of material (99%) that corresponds exactly to the size fragment desired, with a small amount of slightly larger material (as well as the original DNA material).

This section has described PCR in a fairly straightforward way, but the reality is that a successful PCR assay requires significant optimization in the laboratory. The components of the chemical soup, including the concentrations of the DNA starting material, DNA polymerase, primers, and supporting chemicals, are all subject to optimization. The design of the primers is also not necessarily an easy process. Primers must be designed in areas where the DNA sequence is variable, doesn't contain long stretches of a single nucleotide, and has a good mix of all four DNA nucleotides, among other constraints. Moreover, the exact temperature for the primers to optimally anneal to the single-stranded (denatured) DNA requires some guesswork and optimization as well. While the typical characteristics of a PCR assay are well established, the details of a specific PCR assay can require considerable work in the lab for a successful final result.

GENOTYPING WITH RESTRICTION ENZYMES

The intent of the PCR assay is to produce enough DNA copies of the genome region of interest to allow specific analysis of that region. For our purposes, the primary use of PCR is to generate copies of a DNA region containing a polymorphism, which will be studied in many samples so that those samples can be *genotyped.* A secondary use could be to *sequence* the region of DNA (i.e., determine the sequence of letters), which will be discussed later in the chapter. Thus, following the PCR assay, the laboratory technician performs a second type of assay on the newly copied DNA material that will ultimately identify the specific alleles carried by an individual DNA sample at the polymorphism in the copied region.

This section deals with the technique of genotyping through use of **restriction enzymes** (or *restriction endonucleases*). While this is an older method of genotyping, it is still widely used, especially in clinical settings. Restriction enzymes are natural bacterial enzymes that cut or "digest" DNA at very specific sequences. Bacteria use these enzymes as defense mechanisms against invading viruses. Hundreds of restriction enzymes are known to exist, each with its own specific DNA sequence that it recognizes and cuts. As shown in figure 10.4, each restriction enzyme has both a unique DNA recognition sequence and a specific cut site in or near that recognition sequence. Because of their ability to cut DNA at very specific sequences, these unique enzyme machines can be exploited to identify genetic variation in our DNA region of interest. Because polymorphisms are by definition regions of DNA sequence variation, enzymes can be used to identify one variant (i.e.,

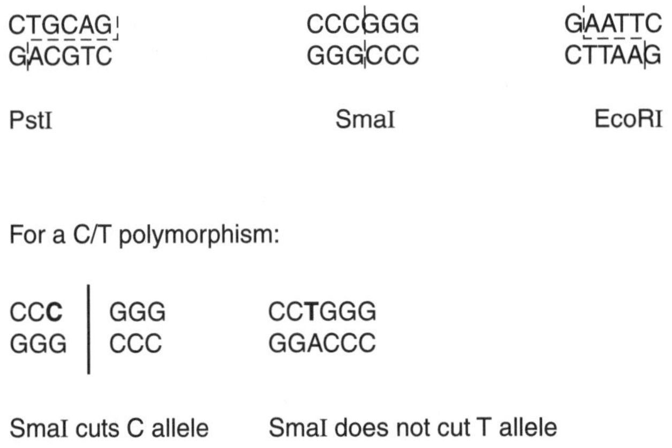

PstI	SmaI	EcoRI
CTGCAG GACGTC	CCCGGG GGGCCC	GAATTC CTTAAG

For a C/T polymorphism:

SmaI cuts C allele	SmaI does not cut T allele
CC**C** | GGG GGG | CCC	CC**T**GGG GGACCC

Figure 10.4 Examples of restriction enzymes and their DNA sequence recognition and cut sites. In the case of a polymorphism, a specific enzyme can be used to cut one of the alleles but not the other allele, and the different lengths of DNA fragments after the cutting can be visualized using gel electrophoresis.

allele) versus another according to whether or not the enzyme cuts the DNA. For example, if a PCR fragment is 200 base pairs in length with a C/T polymorphism present in the center, if a restriction enzyme cuts the C but not the T allele, we can observe different PCR fragment lengths for the two alleles (i.e., two 100 base pair lengths vs. one 200 base pair length for C and T, respectively).

An example of the use of restriction enzymes for genotyping is shown in figure 10.5. Here, we have performed a PCR assay for the region surrounding a C/T polymorphism and can see the PCR fragment sizes after the enzyme digestion for a number of samples (shorter fragments are lower on the image). The sequence immediately surrounding the C/T polymorphism corresponds closely to a recognition sequence for the restriction enzyme EcoRI, such that the presence of the C allele produces the correct recognition sequence (GAATTC) while the presence of the T allele eliminates the recognition sequence (GAATTT). Thus, when the PCR product is exposed to the restriction enzyme, all copies of DNA with the C allele will be cut by the restriction enzyme, and all copies of DNA with the T allele will remain uncut (and thus longer, shown as higher on the image). The restriction enzyme assay, also known as a restriction digest assay, is similar to the PCR assay in that the PCR product is combined in a tube with the restriction enzyme and other supporting chemicals and is exposed to heat (which activates the enzyme) for several hours, allowing the enzyme to find and cut its recognition sequence.

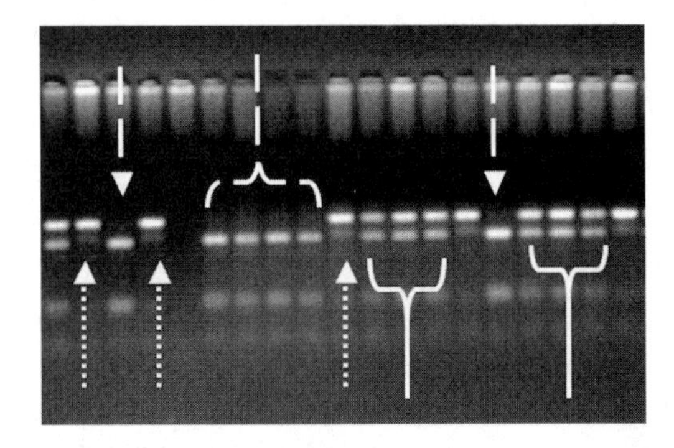

Figure 10.5 An example of a restriction enzyme (or restriction fragment length polymorphism [RFLP]) genotyping assay. The DNA bands were separated by gel electrophoresis, such that the smaller, cut bands are lower on the gel compared to the larger, uncut bands. The absence of either the cut or the uncut band can be used to distinguish homozygotes for each allele. Individuals with two distinct bands are considered heterozygous. In this image, the long-dashed arrows and bracket signify the cut DNA strand; the dotted arrows signify the uncut strand; and the solid arrows and brackets indicate heterozygotes with both cut and uncut strands. Positive control samples are used to verify the genotyping results.

Through use of a process known as **gel electrophoresis,** the DNA from each of our samples can be separated by length, such that shorter segments travel faster than longer segments while exposed to an electrical current in a gelatin-like mold (i.e., the gel, often made with **agarose**). The gelatin-like gel provides resistance for the DNA as it moves toward the opposite side of the gel while following the electrical current. Larger fragments of DNA have more difficulty moving through the gel matrix compared to smaller fragments and will travel through the gel more slowly.

Following PCR and restriction digest of our PCR product, each of the samples is put one by one at the end of the agarose gel and exposed to the electrical current. The DNA travels through the gel once the electrical current is applied, with the length of the DNA determining the speed at which the DNA travels. Because our sample is composed primarily of PCR products of equal length, these products will travel at the same speed through the gel, resulting in a "band" of DNA that can be seen in the gel using various procedures. Once exposed to a restriction enzyme, the PCR product may be cut into two or more smaller fragments, and these smaller fragments will travel faster in the gel in comparison to the uncut PCR product. Thus, as shown in figure 10.5, samples with and without the C allele (cut by the EcoRI enzyme) will appear different from each other because of the presence of the longer and shorter PCR fragments. In this way, homozygotes for each allele and heterozygotes with the C/T genotype can be identified depending on which bands are present in the gel.

KEY POINT

Restriction enzymes, which cut very specific sequences of DNA, can be used to distinguish between two different alleles, resulting in different fragment lengths of DNA that can be visualized by gel electrophoresis. This technique can be used to distinguish the genotype carried for each of many samples at a particular polymorphism.

The selection of a restriction enzyme is dependent on the exact DNA sequence surrounding the polymorphism of interest. With hundreds of restriction enzymes, it is very likely that one will be available to cut the sequence of interest for a given allele without cutting that same sequence when the alternate allele is present. In rare instances, a restriction digest will not be possible and other means of genotyping will be required.

As in all biological laboratory assays, having **positive control samples** is important. In this case, positive control samples would be samples with known genotype that are included with each restriction digest assay to confirm that the enzyme is cutting correctly. Positive control samples can be identified using direct DNA sequencing, described later in the chapter.

Finally, polymorphisms that can be deciphered using these types of restriction enzyme assays are known as **restriction fragment length polymorphisms (RFLPs)** because the length of the PCR product differs depending on the presence or absence of a specific allele (i.e., whether or not the enzyme can cut the sequence). Thus, restriction enzyme assays are often abbreviated as RFLP assays in the research literature. The RFLP assay can be used to identify genotypes for single-nucleotide polymorphisms (SNPs) and insertion/deletions, though not commonly for repeat polymorphisms. In a few instances, PCR alone (without exposure to a restriction enzyme) can be used for genotyping. A common example involves the angiotensin-converting enzyme (ACE) insertion/deletion polymorphism. In this case, the insertion allele is 287 bases in length, which is easily distinguishable via gel electrophoresis from the deletion allele. The PCR product of the deletion allele will be 287 bases shorter than the PCR product containing the insertion allele.

GENOTYPING WITH FLUORESCENT TAGS

While the use of restriction enzymes for genotyping is still widespread, especially for smaller numbers of samples, newer technologies provide faster and easier means of genotyping, particularly for larger numbers of samples. These techniques rely primarily on colorful fluorescent tags that are specific to each of the alleles of interest. Here we discuss this technique in general terms. Multiple varieties of fluorescent genotyping technologies are available, and the purpose of this section is to provide a general introduction to these assays rather than a comprehensive review of different assay types.

Critical to understanding fluorescent technologies is knowing that the nucleotide bases of DNA (A, T, G, and C) can be chemically modified to contain different types of **fluorescent tags.** These tags emit a specific wavelength of light when read in a fluorescent detector, and each nucleotide can be tagged with a unique fluorescent tag, allowing the detector to distinguish among different nucleotides. Recall from the discussion of the PCR assay that a component critical to that assay is the primer, which is a short segment of DNA that is complementary to some of the bases in the DNA sequence of interest. In the case of PCR, primers are designed several nucleotides away from the polymorphism, allowing the production of a PCR product of a few hundred bases in length that can be either sequenced or exposed to a restriction enzyme. In the case of fluorescent techniques, an additional set of primers is designed to bind *at the site* of the polymorphism, for example, with the last base of a primer located at the polymorphism of interest. These primers are known as *allele-specific primers*, or are sometimes referred to as *probes*. Two such primers would be needed for a SNP, one for each allele. Each allele-specific primer would have its own unique fluorescent tag, thus allowing the identification of each of the two alleles. So, when fluorescent tags are used in combination with allele-specific PCR primers, the presence or absence of a particular fluorescent tag can be recognized by the instrument and used to identify the genotype of a DNA sample depending on which tags (i.e., alleles) are present in the sample.

KEY POINT

Fluorescent tags added to allele-specific primers or probe sequences can be used to genotype samples without the use of restriction enzymes.

Figure 10.6 shows an example of using fluorescent tags for genotyping, with allele-specific primers having been designed to bind to the two alleles of a SNP. Two primers are designed, one for each of the two alleles present at the polymorphism, with a different fluorescent tag used for each allele-specific primer. When a specific allele is present (e.g., the A allele), the primer containing that particular complementary nucleotide will bind to that DNA strand, whereas the other primer will not bind. Within the entire sample tube, the binding of primers to A, G, or A and G alleles will alter the fluorescent signal detected by the fluorescent reader, allowing the identification of A/A, G/G, and A/G genotype groups. In this way, homozygotes for each allele are distinguished as having only one fluorescent signal specific to each allele, while heterozygotes are distinguished by having both fluorescent signals present. The fluorescent tags are designed in such a way that they emit their fluorescent light signal only after binding to the correct allele, ensuring accurate allele and genotype identification.

DNA SEQUENCING

In some cases, directly sequencing the nucleotides in a particular DNA fragment is required. This direct DNA sequencing might be needed in the search for positive control samples for genotyping, or could

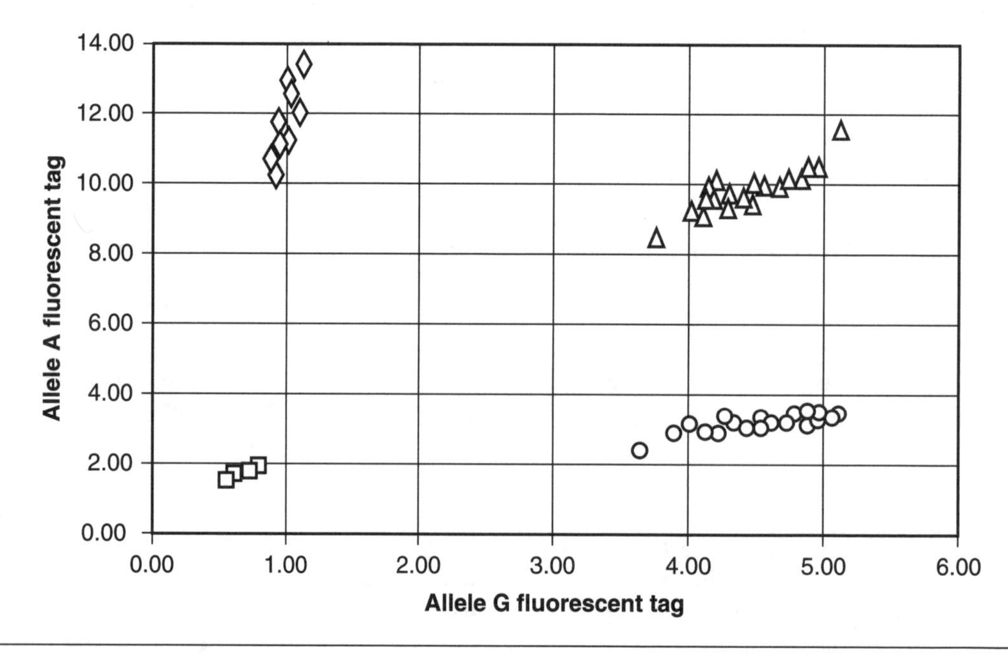

Figure 10.6 An example of the results of a fluorescent tag–based method of genotyping for many individual samples of different genotypes. The diamonds represent the A-allele homozygotes, the circles the G-allele homozygotes, and the triangles the A/G heterozygotes. The squares are either negative control samples or failed samples. In the lab, these shapes would also be color coordinated to assist with visual inspection of the four groupings.

SPECIAL FOCUS
DNA Microarrays

With the millions of SNPs of interest within the genome, researchers are often reluctant to prioritize one as more important than another, especially in the exploratory phases of a genetic study. Thus, "high-throughput" technologies have been developed to allow many thousands of polymorphisms to be genotyped in a DNA sample within a single assay. The result is an incredible amount of genetic information for a single individual, which can then be used to help identify potential genes and alleles of interest for a particular phenotype.

The most common of these high-throughput genotyping techniques is known as the DNA SNP **microarray** (often called "SNP chips"). In PCR, complementary primers are designed that are introduced to the DNA to allow copying of a specific region of DNA. In the case of microarrays, the sample DNA is instead exposed to thousands of different complementary primers that are attached to a small glass surface (the microarray). Similar to what occurs with PCR, each primer has been designed to bind to a specific region of the DNA, including specific alleles within polymorphisms (i.e., allele-specific primers). Each type of primer is bound to a very small spot on the array, allowing many thousands of different primers to fit within the relatively small area of a microarray (often only a few centimeters square in total area). The sample DNA is then introduced to the array, and complementary DNA segments will bind to complementary primers. For example, for an A/G polymorphism in a particular gene, the microarray will have primers specific to both the A and the G alleles of that DNA segment. If a person is homozygous for the A allele (i.e., A/A genotype), then that part of the person's DNA will bind only to the primer containing the base complementary for the A allele. Fluorescent tags are included with the DNA to allow recognition of which primers are bound. Thus, similar to the use of fluorescent tags for genotyping a PCR product, the fluorescence is measured for both sets of primers and the combination of fluorescent tags is used to identify the genotype. The unique attribute of the microarray is that this process is performed several thousand times in a single assay, with primers for several thousand SNPs present on the microarray. The microarray instrument recognizes the specific, microscopic location of each primer, allowing a computer to distinguish the specific fluorescent signals corresponding to each set of primers, which are each specific for a different polymorphism. An amazing amount of information can be obtained for a single DNA sample in a single assay. These SNP chips will likely have importance for *genome-wide association studies,* as discussed in chapter 7.

Similar microarrays have been used for several years for the analysis of mRNA from a tissue sample for studies of gene expression. Using the same idea of bound, sequence-specific primers, mRNA strands from a tissue sample are exposed to the microarray. Messenger RNA sequences that match a particular primer sequence on the microarray will bind to that primer, and the amount of mRNA that binds (as indicated by a fluorescent tag) is proportional to the amount of mRNA within the tissue sample. Remarkably, thousands of matching primers can be bound to the single small location on the microarray. Thus, each spot on the microarray can be read by the instrument to determine both whether a specific mRNA product was present in a tissue and how much of that product was present. This gives researchers (and soon, clinicians) a window on the process of gene expression within the tissue—the extent to which a gene is being transcribed under certain conditions (e.g., before and after some therapeutic intervention).

The use of microarray technology has exploded in biological and genetic research. The result has been an incredible increase in the amount of biological data obtained from a tissue or DNA sample. What is often referred to as the "mountain" of resulting data can be a serious challenge from a data management and statistical analysis standpoint. **Bioinformatics** is an emerging computer science and statistics field dedicated to the analysis of such large biological databases.

be used to search for polymorphisms among several individual samples. Stated simply, **DNA sequencing** is the process of identifying the specific nucleotides and their order within the DNA segment—in effect, knowing the spelling of the letters in the DNA strand. The Human Genome Project was an immense effort in DNA sequencing, such that all 3.1 billion letters of DNA were fully sequenced and put into order with 99.99% accuracy.

DNA sequencing also relies on the use of fluorescent tags, as discussed earlier. A PCR reaction is performed as previously described, forming many copies of the DNA region of interest. A second *sequencing reaction* is then performed with the PCR product, though only one PCR primer is used so that only one strand of the DNA double-stranded molecule is sequenced in a particular assay. The goal with this second PCR-like reaction is to make additional copies of each of the PCR products, but with the incorporation of fluorescent tags in order to determine the DNA sequence. Nucleotides modified to contain the fluorescent tags are added to the chemical soup of the sequencing reaction. In the case of DNA sequencing, these tagged nucleotides are included with untagged nucleotides such that they are present as only a small fraction of the total nucleotides in the PCR cocktail. This is critical, because the tagged nucleotides, known as dideoxyribonucleic acids (dA, dG, dC, and dT, vs. the typical A, G, C, and T), prevent the progress of the polymerase enzyme in copying DNA during PCR. In other words, once a tagged nucleotide is added to the growing DNA strand during copying, the polymerase stops and no additional nucleotides can be added. It is this unique feature of tagged nucleotides that is critical to the process of DNA sequencing.

Because the tagged nucleotides are added as just a small fraction of the total number of nucleotides in the PCR cocktail, the polymerase enzyme is only rarely stopped by the presence of a tagged nucleotide. In other words, the sequencing reaction and DNA copying continue almost as in a standard PCR assay, but the DNA copying is randomly stopped during sequencing by the occasional introduction of a tagged base. In the millions of copies of PCR products produced in a sequencing assay, at random some small number of products will have been produced with the very *first* base of the DNA segment fluorescently tagged, some products with the *second* base tagged, some products with the *third* base tagged, and so on, with some products having only the *last* base tagged. In this way, every base within the DNA segment will have been tagged in some number of PCR products in the sequencing assay.

Recall from the discussion of gel electrophoresis that DNA fragments exposed to an electrical current in a gel will move according to their length, such that the smallest fragments will travel faster and the longest fragments more slowly. In the case of a DNA sequencing assay, every possible PCR fragment length is produced, with the last base of each fragment being the fluorescently tagged nucleotide that stopped the copying process. So, once the fragments are ordered from shortest to longest by the electric current, the final tagged nucleotide can be identified by the fluorescent instrument reader for each PCR fragment. The DNA sequence can simply be read as the order of tagged bases that travel through the gel, shortest to longest. Polymerase chain reaction products of up to about 700 bases can be accurately sequenced using commonly available technologies. The method of DNA sequencing described here is known as the *chain termination* or *Sanger method*, developed by Dr. Frederick Sanger in 1975. Dr. Sanger received his *second* Nobel Prize in Chemistry for this work in 1980 (the first was for determining the amino acid sequence of insulin, awarded in 1958).

Once the tagged PCR fragments are ordered from shortest to longest and the fluorescent reader identifies the tagged nucleotides, an image or **chromatogram** of the DNA sequence is generated. Figure 10.7 shows part of a chromatogram of DNA sequence for three individual samples. Notice how all of the samples show characteristic wave patterns, with each nucleotide shown as a single wave. Multiple waves (nucleotides) are shown in series, reflecting the sequence of nucleotides in the DNA sequence. In the laboratory, each waveform would have a distinct color, specific to each nucleotide; here, we've redrawn the image to work without the benefit of color. Notice that the order of the nucleotides is the same across all of the samples, reflecting the fact that the vast majority of nucleotides in the genome are shared by all individuals. Notice also in position 5 from the left in the sequence that the waveform pattern differs among the individual samples. Here, we see that one individual has a distinct G wave (shown as black), one individual has a distinct A wave (shown as dashed), and one individual has two smaller G and A waves (identified as R in figure

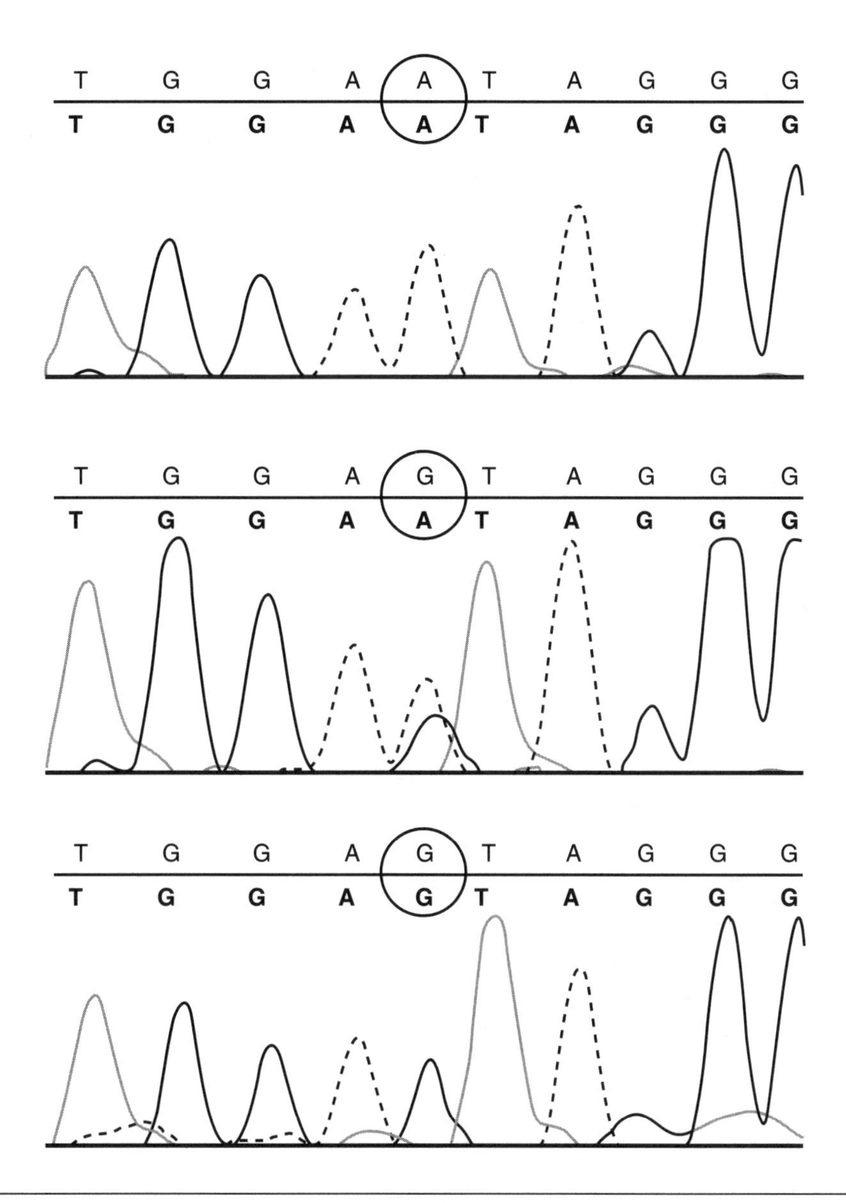

Figure 10.7 DNA sequencing chromatograms for a segment of DNA for three different individuals. The short segment shown here is used to demonstrate how polymorphisms can be identified within the DNA sequence, as evidenced by the A/A, A/G, and G/G genotypes shown for each of the three chromatograms, respectively. The top line of sequence reflects the true sequence, as interpreted by the researcher, while the second line of sequence reflects the computer software's interpretation of the sequence. In the laboratory, the black line would be black, representing G nucleotides; the dashed line would be green, representing A nucleotides; and the gray line would be red, representing T nucleotides; C nucleotides would be blue, but none are present in this sequence. Note also that "noise" in the chromatogram must be interpreted as being real or as simply background levels of fluorescence that should be ignored when the DNA sequence is being determined.

10.7) in the same position. This is the characteristic chromatogram sequence of a SNP, for a G-allele homozygote, an A-allele homozygote, and a G/A heterozygote, respectively. Thus, DNA sequencing can be used both to identify genotypes for a particular polymorphism for choosing positive control samples and to search for novel polymorphisms among DNA samples from several individuals. DNA sequencing is considerably more expensive than the other genotyping techniques described, so it is not commonly used as a genotyping tool for large numbers of samples. The exception is for genotyping repeat polymorphisms, which often cannot be genotyped using RFLP or the fluorescent tag methods described earlier.

SUMMARY

To perform a genetic study in the research lab or in the clinic, the first requirement is a DNA sample. This is most commonly obtained from the white blood cells of a whole-blood sample, though cheek cells can be used as an alternative source. Once the DNA is extracted, it becomes the target of PCR, which amplifies or copies the particular DNA region of interest within the genome many, many times. This PCR product can then be used in a variety of assays, including genotyping by restriction enzymes, genotyping with allele-specific primers containing unique fluorescent tags, microarrays, or direct DNA sequencing. The genotyping methods allow identification of the specific alleles carried by an individual at a particular polymorphism, while DNA sequencing is used to determine the exact nucleotides and their order within a small region of DNA sequence.

KEY TERMS

agarose
annealing
bioinformatics
buccal cell
buffy coat
chromatogram
denaturation
DNA extraction
DNA polymerase
DNA primers
DNA sequencing
elongation
extension
fluorescent tag
gel electrophoresis
microarray
polymerase chain reaction (PCR)
restriction enzyme
restriction fragment length polymorphism (RFLP)

REVIEW QUESTIONS

1. When taking a blood sample is not possible, what is another source of DNA for performing a research study or clinical test?
2. What are the three steps of a PCR assay? Describe the major events occurring in each of those steps. In general terms, describe the end result of the PCR assay.
3. What is a restriction enzyme, and how can it be used for genotyping a polymorphism after the PCR assay?
4. How are fluorescent tags used for genotyping?
5. What are the primary uses for DNA sequencing? How can a polymorphism be identified in a chromatogram sequence?

PART

III

CURRENT FINDINGS AND EXTENSIONS OF GENETICS RESEARCH

In part III of the book, we address some of the current research findings in the area of genetics in exercise and health as well as discuss some of the current issues in the field. For an author, this task is challenging, because last year's research findings are hardly "current," and the important issues within a field are often quick to change. The challenge is to present meaningful information that is not immediately outdated once a book is published. Chapter 11 provides an overview of current research findings for a number of different genes, with the goal of giving readers concrete examples of how genetics has already been applied to health and fitness research questions. The remainder of part III provides a broader overview of various issues in the wider field of genetics and health. Chapter 12 discusses the latest trends and applications of genetics in health and medicine, known as personalized medicine. Finally, the ethical challenges of genetics in both society and sport are discussed in chapter 13. My goal in the final chapters of part III is to introduce readers to these issues, providing the foundation necessary for moving forward and adapting in this emerging discipline.

CHAPTER

11

CURRENT RESEARCH FINDINGS IN THE GENETICS OF EXERCISE SCIENCE AND HEALTH

At this point in the book, we've covered the bulk of the basic information needed for an understanding of genetic association studies. In this chapter, we put those skills into practice by examining some concrete examples of genetic associations with health and fitness phenotypes. In this quickly moving field of human genetics, an update on current research is a nearly impossible task, as whatever is written is rapidly outdated by ongoing research. So, the goal of this chapter is to provide brief examples of some key findings in order for readers to see the typical progression of genetic association research and see the types of questions that are of current interest to researchers in the field. As you read these sections, consider searching for recently published literature in each area to see how the findings have changed since 2006, when these sections were written.

OVERVIEW OF CURRENT FINDINGS

The progress of research in human genetics as it relates to health and fitness phenotypes has been mixed. While several research groups are working on identifying genes and alleles important for a variety of phenotypes, few genes have been conclusively identified as key contributors to these traits. Progress in the field can be measured by the explosion of genetic association studies published since 2000, about the time the human genome sequence began to be publicly available. By the year 2000, for example, 20 papers had been published examining genetics and endurance performance, two papers on genetics and strength or anaerobic power traits, and eight papers looking at genetics in relation to the response of blood lipids and inflammatory markers to exercise training. But, by the end of 2005, the published literature had expanded considerably, with the number of papers in each of those areas increased to 53, 23, and 32, respectively, and those numbers continue to grow. This growth is also emphasized in the Special Focus section of the chapter.

Four examples of genetic associations with health- and fitness-related phenotypes have been selected for discussion in the current chapter. Certainly, other examples could have been highlighted (as discussed in the Special Focus section), but these four examples represent a nice spectrum of the types of questions being asked.

The first example is the research on a nonsense mutation in a skeletal muscle gene known as ACTN3, which is present in type II muscle fibers. While the nonsense mutation completely disables the ACTN3 protein, there is no apparent muscle disease in men and women homozygous for the mutation. Researchers are now focused on whether the mutation has an influence on nondisease muscle phenotypes.

The second example is related to Alzheimer's disease. Development and progression of the disease are associated with genetic variation in the ApoE gene. With this strong association as a foundation, researchers moved to understanding the interaction of ApoE genotype with various environmental factors in the development of the disease. In this chapter we focus on the research aimed at studying the interaction of physical activity with ApoE genotype.

The third example is from the exercise performance world. What makes an elite athlete—genetic factors, exceptional motivation, optimal training techniques, or all of the above? This is a question that has driven researchers to identify specific genetic factors underlying exercise performance. Here, we focus on the considerable work targeting sequence variation in the ACE gene and different types of exercise and sport performance.

Finally, we discuss the research surrounding the myostatin gene and its possible relevance to variation in skeletal muscle phenotypes, particularly muscle mass. This example fits most closely with that of the progression discussed in chapters 8 and 9, in that a candidate gene was identified from research in animal models, genetic variation was identified and prioritized, and various genetic association studies were performed to test hypotheses regarding the genetic variation and muscle phenotypes.

For all of these examples, key citations are included that provide a starting point for readers interested in learning more about the given topic. References for this chapter are cited in these key citation lists rather than in the bibliography. The citation lists are not complete, but represent important papers useful for gaining insight into valuable background information and key findings for each area.

SPECIAL FOCUS

The Human Gene Map for Performance and Health-Related Fitness Phenotypes

In an effort to provide researchers with a quick method of gaining insight into the current research findings in the areas of exercise performance and health-related fitness, a group of researchers began a project to completely review the published findings for genetic associations for these traits. Obviously, such a task is not easy; multiple phenotypes fit within the broad areas of performance and health-related fitness, not to mention the various exercise stimuli that could be studied in combination with the various traits. Nonetheless, the research group, led by Dr. Claude Bouchard, provided a comprehensive review of the related literature (i.e., all such literature published until the end of 2000). Moreover, the intent of the group remains to publish a regular update of the review, with all new papers described and included in each update. The result, titled, "The Human Gene Map for Performance and Health-Related Fitness Phenotypes," was first published in *Medicine and Science in Sports and Exercise* in 2001, and has been updated nearly every year since then in the same journal.

As shown in figure 11.1, the review provides a graphical map of all of the genes and gene regions that have been positively associated with performance and health-related fitness traits. That map is updated yearly, as are the extensive tables within the review, which list those studies with positive genetic association or linkage analysis results for various phenotypes. Studies reporting negative associations are reviewed in the paper in the year following their publication, but those genes are not included in the map and the papers are not maintained in the review beyond that year. The goal of the first review was to provide readers with a history of the genetic studies done within the field until the end of 2000; the additions to the map provide a compendium of the most recent research across the broad range of phenotypes within the performance and health-related fitness areas. From the most recent gene map available for discussion in this book (Rankinen, Bray, et al., 2006), 170 genes and gene regions have been positively associated with a trait of interest. That is considerable growth from the 29 genes and gene regions reported in the first version of the map in 2001! Expect further growth in future updates of the review.

(continued)

(continued)

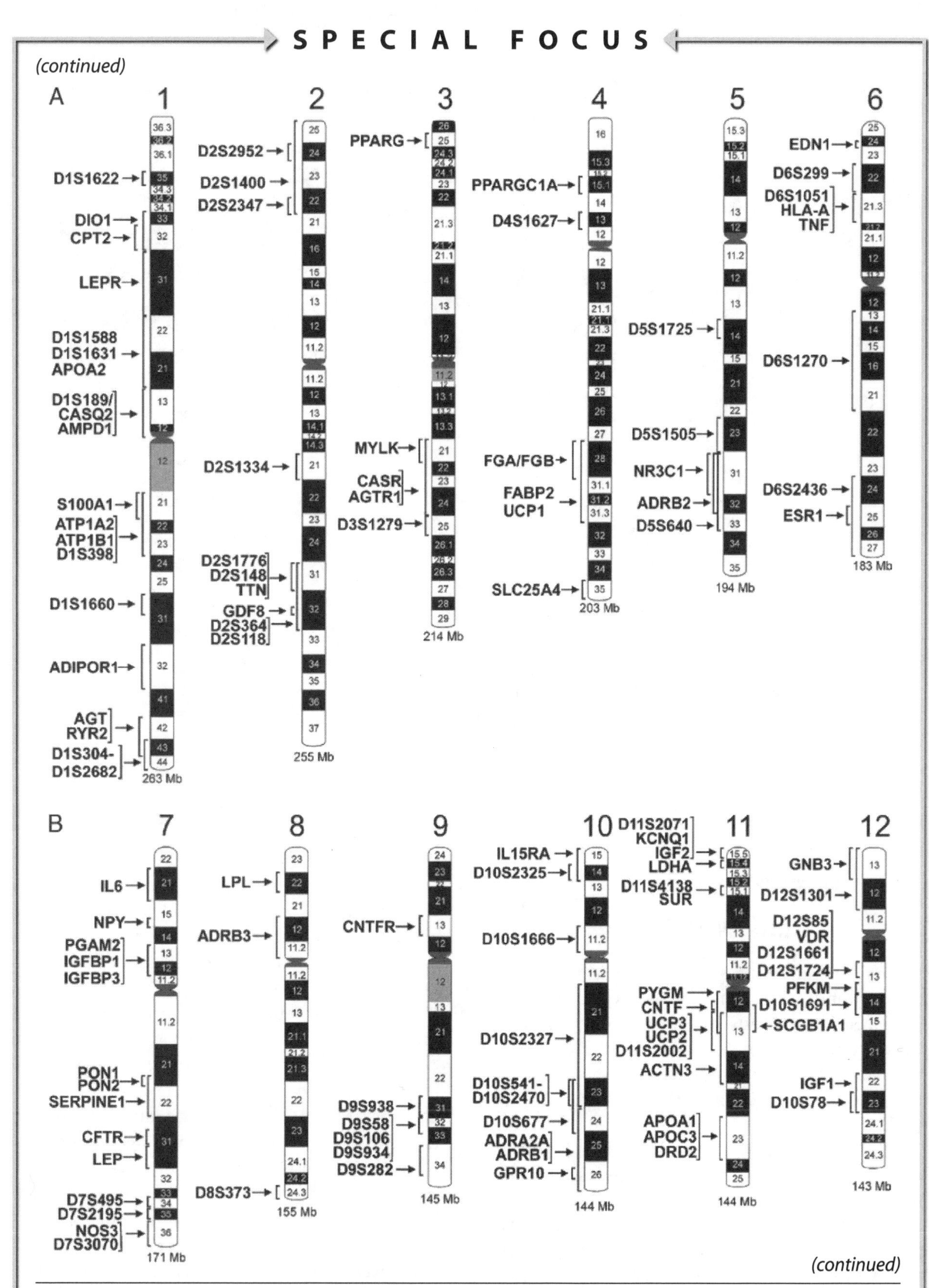

(continued)

Figure 11.1 The genes and gene regions across all of the human chromosomes that have positive associations with performance or health-related fitness phenotypes. The figure is from the 2004 update of the gene map review article published in 2005. Subsequent updates of the gene map have been published.

Reprinted, by permission, from B. Wolfarth et al., 2005, "The human gene map for performance and health-related fitness phenotypes: The 2004 update," *Med. Sci. Sports Exerc.* 37(6): 881-903.

SPECIAL FOCUS

(continued)

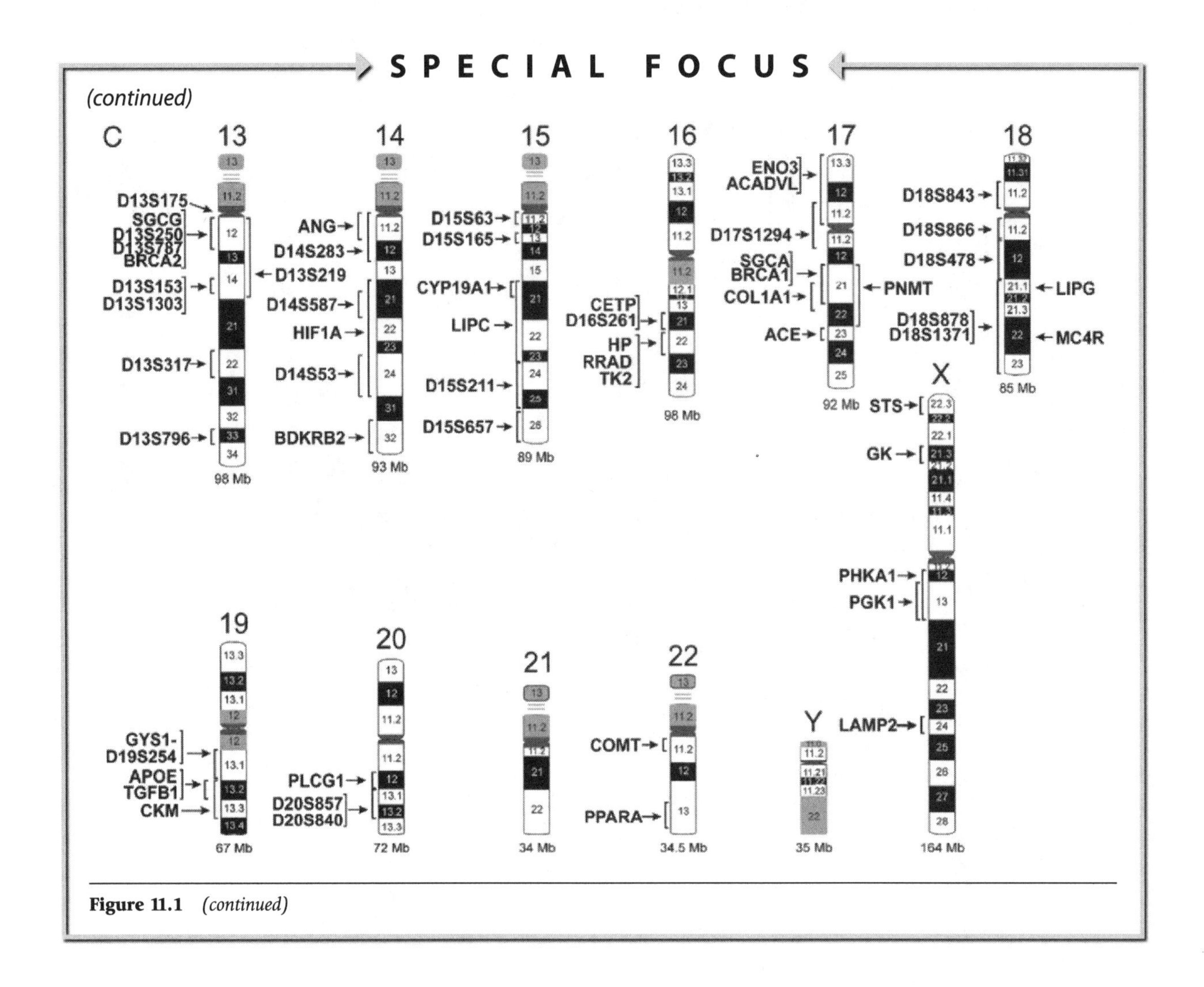

Figure 11.1 *(continued)*

ACTN3 AND MUSCLE PERFORMANCE

In some cases in genetics research, a newly identified polymorphism in a gene can be hypothesized to influence various traits, the gene or polymorphism thus providing the foundation for the design of a genetic association study. Such is the case with the gene described in this section: alpha-actinin 3 (ACTN3).

ACTN3 is one of a family of alpha-actinin proteins that are important in a number of tissues. Two proteins of specific interest for us are ACTN2 and ACTN3, both of which are expressed in skeletal muscle cells. Skeletal muscles are made up of multiple muscle cells, which are grouped as type I or type II according to various metabolic and contractile characteristics. Type I fibers rely predominantly on aerobic metabolism and have lower contractile forces than type II fibers, which have higher contractile forces and rely predominantly on anaerobic metabolism. ACTN2 is expressed in all fiber types of skeletal muscle, while ACTN3 is found only in type II fibers.

In 1999, North and colleagues identified a nonsense polymorphism in the ACTN3 gene, known as R577X (arginine replaced by a stop codon at codon number 577). Carriers of the X/X genotype show complete absence of ACTN3 protein in their muscle fibers. In the mid-1990s, various studies of muscular disorders had identified individuals completely deficient for ACTN3 protein, hinting that perhaps this deficiency was related to the development of muscle disease, specifically certain types of muscular dystrophy. Thus, the R577X mutation could be useful in explaining muscle disease, which makes intuitive sense given that the mutation is a nonsense allele.

To the contrary, North and colleagues found that the X allele is quite common in several different populations (X-allele frequency ranging from 25% to 50%), meaning that many people around the world (X/X genotype carriers) are deficient for ACTN3 protein. The association of ACTN3 deficiency with muscular disorders was a chance finding and in fact ACTN3 deficiency does not appear to have harmful effects on muscle function. Since ACTN2 is also expressed in type II fibers, ACTN3 deficiency is apparently not deleterious because of functional redundancy between the two proteins.

So if the complete lack of ACTN3 is not associated with muscle disease or any other apparent health problems, does it play a role in a less severe muscle phenotype, perhaps explaining typical differences among individuals for muscle function? North and colleagues hypothesized that ACTN3 genotype might be a genetic factor important to normal phenotype variation in muscle function, especially given its specific location in type II muscle fibers. In other words, does the lack of ACTN3 protein in type II muscle fibers in X/X individuals affect muscular performance? The group sought to answer this question by performing a case–control study of elite athletes and control subjects.

This follow-up study, published by Yang and colleagues in 2003, examined ACTN3 genotype in 436 subjects of Northern European (Caucasian) ancestry, half of whom were elite Australian athletes from a variety of sports. The athletes were grouped according to sprint or power-related sports (e.g., short-distance sprinters, swimmers, and cyclists) or endurance sports (e.g., long-distance runners, swimmers, cyclists, and rowers). The researchers hypothesized that the X/X genotype (ACTN3 deficiency) would be a disadvantage in sprint or power-related sports, in which type II fibers are recruited, so fewer X/X genotypes would be observed in the sprint or power athletes compared to controls and endurance athletes. As shown in figure 11.2, reprinted from their paper, that is exactly what Yang and coworkers observed. The frequency of the X allele was significantly lower in sprint or power athletes (6%) compared to both controls (18%) and endurance athletes (24%). In fact, the X/X genotype was completely absent in both female sprint or power athletes and male and female Olympic sprint or power athletes. The frequency of the R/R genotype was higher in the sprint or power

> **KEY POINT**
>
> The R577X polymorphism in the ACTN3 gene results in the complete absence of alpha-actinin 3 protein in X/X genotype carriers. Though these individuals do not have overt muscle disease, the complete lack of this type II muscle fiber protein may affect muscle performance or other muscle-related phenotypes in these individuals.

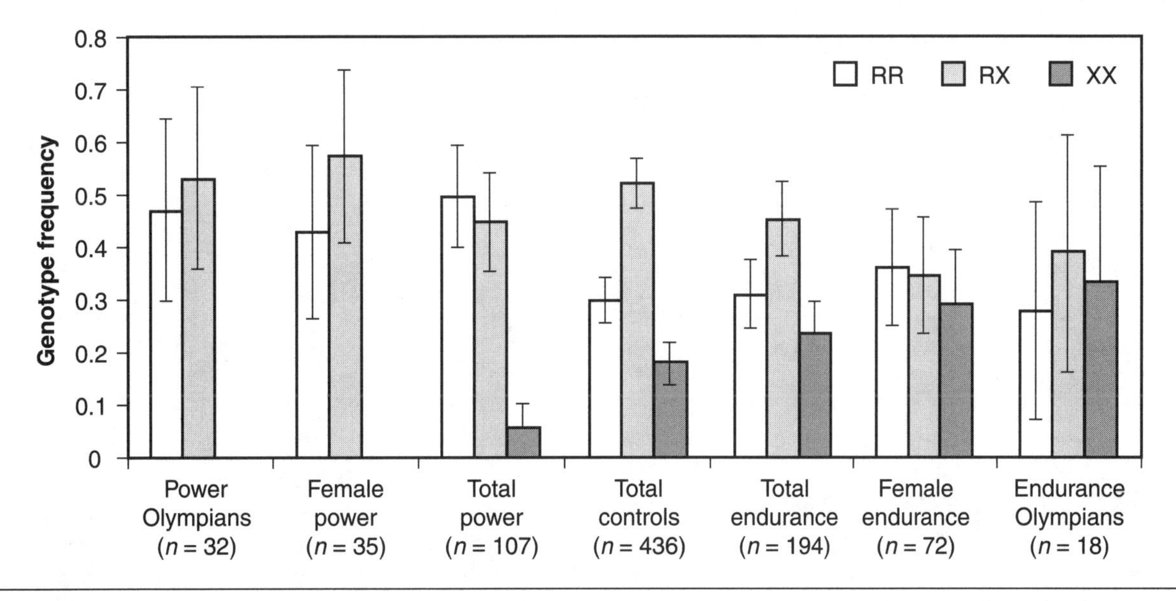

Figure 11.2 Genotype frequencies for the R577X polymorphism within the ACTN3 gene in various groups. The middle group is the control group, showing an X/X genotype frequency of 18%. The groups to the left of the control group are the sprint or power athletes; notice the low frequency or complete absence of the X/X genotype in these groups. Conversely, the groups shown to the right of the control group are the endurance athletes, in whom the X/X genotype is at least as frequent as in the controls.

Reprinted from N. Yang et al., 2003, "ACTN3 genotype is associated with human elite athletic performance," *Am. J. Hum. Genet.* 73(3): 627-631, by permission of the University of Chicago Press.

athletes than in the control or endurance athlete groups as well. Thus, the authors concluded that the R allele of the ACTN3 R577X polymorphism provides an advantage for sprint or power-related activities compared to the X allele.

From the results of Yang and colleagues, several questions can be generated: How does muscle performance differ among untrained men and women of different ACTN3 genotypes (i.e., subjects without an athletic background)? Is the influence of ACTN3 genotype on muscle performance seen only after extensive exercise training? Are other muscle function variables (e.g., strength, fiber type proportions, activities of daily living) influenced by ACTN3 genotype, especially in older individuals, who often show declines in muscle performance? The paper by Yang and colleagues provides for several new hypotheses that will need to be tested for a full understanding of the implications of the X/X genotype. Because the X/X genotype completely disables the ACTN3 protein, the gene is an important candidate for muscle phenotypes since the allele is known to be functional.

Other research groups are working quickly to follow up on the original ACTN3 studies. Niemi and Majamaa (2005) reported findings very similar to those of Yang and colleagues, showing lower X/X genotype frequencies in Finnish sprint athletes compared to endurance athletes. In fact, none of the top sprint athletes carried the X/X genotype, a result that was similar to the findings of Yang and colleagues. Also in 2005, Clarkson and coworkers reported on the influence of ACTN3 genotype on several muscle phenotypes. In their paper, upper arm strength was compared across ACTN3 genotype in men and women before and after a strength training program. This was a prospective genetic association study, with subjects recruited regardless of genotype, though the subjects had to be healthy and sedentary in order to participate. In this study, women carriers of the X/X genotype had lower baseline isometric muscle strength than other genotypes, as might be expected from the paper by Yang and colleagues; but contrary to expectations, the X/X group showed greater strength increases in response to strength training compared to the R-allele carriers. In fact, ACTN3 genotype was not associated with either baseline or after-training muscle function in men. Despite the very clear evidence for a role for ACTN3 genotype in muscle function in Yang and colleagues' paper, the follow-up study by Clarkson and colleagues did not lend support to that hypothesis.

At the time of this writing, there is no clear consensus on the meaning of these inconsistent results, but certainly additional papers can be expected from a number of groups that should help shed light on the influence of ACTN3 genotype on muscle function.

APOE AND COGNITIVE IMPAIRMENT

Alzheimer's disease, which is a progressive and currently incurable age-related dementia, is a leading cause of death in the United States, with considerable associated health care costs. As the proportion of older individuals increases over the next few decades, the number of people with Alzheimer's disease is expected to increase dramatically, as are the associated health care costs. Fortunately, the disease is

Citations for ACTN3 and Muscle Performance

Clarkson, P.M., J.M. Devaney, H. Gordish-Dressman, P.D. Thompson, M.J. Hubal, M. Urso, T.B. Price, T.J. Angelopoulos, P.M. Gordon, N.M. Moyna, L.S. Pescatello, P.S. Visich, R.F. Zoeller, R.L. Seip, E.P. Hoffman. ACTN3 genotype is associated with increases in muscle strength in response to resistance training in women. *J. Appl. Physiol.* 99: 154-163, 2005.

Mills, M.A., N. Yang, R.P. Weinberger, D.L. Vander Woude, A.H. Beggs, S. Easteal, K.N. North. Differential expression of the actin-binding proteins, a-actinin-2 and -3, in different species: Implications for the evolution of functional redundancy. *Hum. Mol. Genet.* 10: 1335-1346, 2001.

Niemi, A.K., K. Majammaa. Mitochondrial DNA and ACTN3 genotypes in Finnish elite endurance and spring atheltes. *Eur. J. Hum. Genetic.* 13: 965-969, 2005.

North, K.N., N. Yang, D. Wattanasirishaigoon, M.A. Mills, S. Easteal, A.H. Beggs. A common nonsense mutation results in a-actinin-3 deficiency in the general population. *Nature Genet.* 21: 353-354, 1999.

Yang, N., D.G. MacArthur, J.P. Gulbin, A.G. Hahn, A.H. Beggs, S. Easteal, K.N. North. ACTN3 genotype is associated with human elite athletic performance. *Am. J. Hum. Genet.* 73: 627-631, 2003.

progressive, so individuals do not suddenly acquire the disease in its devastating later stages; rather, it develops over time, progressing from mild cognitive impairment to severe disease over a number of years. Thus, the potential exists for slowing the progression of the disease through various treatments, including medications, physical activity, diet, and other behavioral interventions.

There are two forms of Alzheimer's disease, early onset and late onset. Early-onset Alzheimer's disease appears to be predominantly genetic in nature, and several rare but high-risk gene mutations have been associated with the disease. Early-onset patients account for only a small fraction of total Alzheimer's disease cases. The vast majority of affected patients have late-onset Alzheimer's disease, occurring most commonly after the age of 65 years. Late-onset Alzheimer's disease is a heritable disorder, in that it aggregates in families and has been shown to have a genetic component, with heritability (h^2) values as high as 60%. After heritability was determined for the disease, studies were performed in families to identify areas of the genome that might be carrying a gene important for the disease; these studies are known as linkage analyses. Specific genes are not identified in these studies, but candidate gene lists can be generated from the regions that are identified as being important for a phenotype or disease trait (see chapter 7 for a detailed discussion of linkage analysis).

In 1993 in a large linkage study, Strittmatter and colleagues attempted to identify the specific gene located within a genome region on chromosome 19 that was associated with late-onset Alzheimer's disease. Their work revealed an unlikely candidate as an explanatory gene: apolipoprotein E (ApoE). The ApoE protein, rather than being known for importance in brain physiology, is a component of very low-density lipoproteins (VLDL) in the blood and plays an important role in lipoprotein metabolism. Nonetheless, the group identified a strong association between genetic variation in the ApoE gene and Alzheimer's disease.

Two neighboring missense single-nucleotide polymorphisms (SNPs) in the ApoE gene result in three common haplotypes present in humans, known as ε2, ε3, and ε4, each carrying a unique combination of alleles at the two missense polymorphisms. In other words, this set of SNPs is in a haplotype block, with each haplotype labeled as ε2, ε3, and ε4, and the possible diplotypes labeled as ε2/ε2, ε2/ε3, ε3/ε3, ε3/ε4, and ε4/ε4. In this case, having just two SNPs allows the direct measurement of diplotype (see discussion in chapter 7). For the purposes of clarity, and for consistency with the ApoE literature, we will refer to these haplotypes as alleles and the diplotypes as genotypes.

The ApoE ε3 allele is the most common allele and the ε2 allele the least common in most populations, with the ε4 allele intermediate in frequency. In a case–control study by Strittmatter and colleagues, the frequency of the ε4 allele was 50% in Alzheimer's disease patients compared to only 16% in an age-matched control population. This positive association between ApoE genotype and Alzheimer's disease risk was confirmed in a number of follow-up studies by that and other research groups across the world. The presence of the ε4 allele increases risk of Alzheimer's disease by three- to fourfold in heterozygotes (i.e., ε3/ε4 genotypes) and by eightfold in ε4/ε4 homozygotes, and is present in about half of Alzheimer's disease patients. Moreover, the age of onset of disease is lower in ε4 carriers, especially for ε4 homozygotes. How the ApoE gene and the ε4 allele in particular result in increased Alzheimer's disease risk is unclear at present, though several hypotheses are being tested.

KEY POINT

The ApoE ε4 allele is a strong predictor of risk for Alzheimer's disease, especially in homozygote carriers. Researchers are now moving to address potential modifiers of this genetic association, in particular asking if physical activity can modify the heightened risk of cognitive dysfunction associated with the ε4 allele.

From a health care perspective, can Alzheimer's disease risk be reduced in ε4-allele carriers using various means? Here, we will focus on physical activity as a possible intervention. Exercise training has been shown to improve general cognitive functioning in older adults, so the question of an ApoE gene × physical activity interaction was specifically addressed in 2001 by Schuit and coworkers. In a retrospective association study of 347 older Dutch men, Schuit and colleagues examined cognitive function over a three-year period in relation to both ApoE genotype and physical activity levels. As shown in figure 11.3, the results were quite remarkable. As expected, ApoE genotype was a strong predictor of declines in

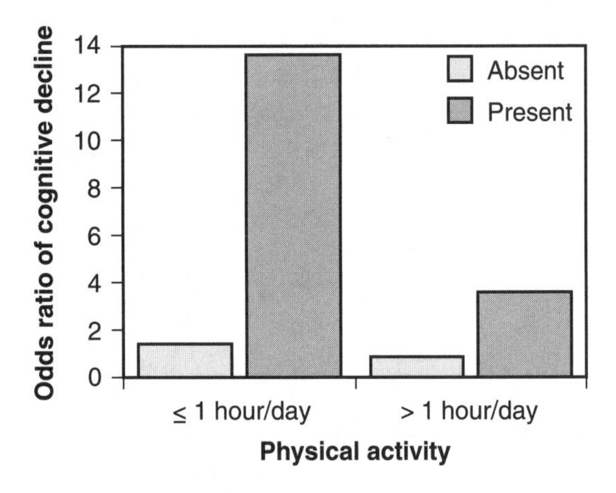

Figure 11.3 The risk (known as the adjusted odds ratio) for cognitive decline on the vertical axis for four groups of subjects distinguished by ApoE genotype and physical activity levels. Those individuals who are noncarriers for the ϵ4 allele, represented by the light bars, have similar risk for the development of cognitive decline regardless of physical activity levels. Carriers of the ϵ4 allele (dark bars), on the contrary, show elevated risk of cognitive decline, but that risk is considerably reduced in physically active ϵ4-allele carriers.

Reprinted, by permission, from A.J. Schuit et al., 2001, "Physical activity and cognitive decline, the role of the apolipoprotein e4 allele," *Med. Sci. Sports Exerc.* 33(5): 772-777.

cognitive function (dark bars in the figure). But in those ϵ4-carrying men who maintained high levels of physical activity, the risk of cognitive decline was reduced dramatically. In fact, sedentary ϵ4 carriers had a 13.7-fold higher risk of cognitive decline compared to physically active noncarriers of the ϵ4 allele (the reference group), while the risk for physically active ϵ4 carriers was only threefold higher than that for the reference group. Thus, a strong interaction was shown, such that the presence of high levels of physical activity significantly reduced the disease risk of the ϵ4 allele, hence slowing the progression of cognitive decline in this group.

These results have been followed up in at least two additional studies. In 2005, Podewils and colleagues reported a study of over 3000 older men and women who were followed over a five-year period with measures of physical activity, genotype, and the presence of dementia. In that study, more physically active subjects showed lower risk of developing dementia compared to sedentary individuals, as expected; however, the results did not show an interaction with ApoE genotype. Contrary to the findings of Schuit and colleagues, carriers of the ϵ4 allele appeared to develop dementia at similar rates regardless of physical activity levels.

In late 2005 a third paper, by Rovio and colleagues, provided additional support for the results of Schuit and colleagues. Rovio's group examined the associations of physical activity and ApoE genotype with risk of dementia and Alzheimer's disease in 1449 individuals followed over a 21-year period. The authors reported that physical activity was associated with risk of dementia, with a 50% lower risk of onset of dementia and Alzheimer's disease in physically active compared to sedentary subjects. When ApoE genotype was considered in addition to physical activity, they found that ApoE ϵ4 carriers who were physically active had an even greater reduction in risk compared to sedentary ϵ4 carriers. In fact, the risk of dementia for physically active ϵ4 carriers was equal to that of

Citations for ApoE and Cognitive Impairment

Clark, C.M., J.H.T. Karlawish. Alzheimer disease: Current concepts and emerging diagnostic and therapeutic strategies. *Ann. Intern. Med.* 138: 400-410, 2003.

Kramer, A.F., L. Bherer, S.J. Colcombe, W. Dong, W.T. Greenough. Environmental influences on cognitive and brain plasticity during aging. *J. Gerontol. Med. Sci.* 59A: 940-957, 2004.

Podewils, L.J., E. Guallar, L.H. Kuller, L.P. Fried, O.L. Lopez, M. Carlson, C.G. Lyketsos. Physical activity, ApoE genotype, and dementia risk: Findings from the cardiovascular health cognition study. *Am. J. Epidemiol.* 161: 639-651, 2005.

Rovio, S., I. Kareholt, E-L. Helkala, M. Viitanen, B. Winblad, J. Tuomilehto, H. Soininen, A. Nissinen, M. Kivipelto. Leisure-time physical activity at midlife and the risk of dementia and Alzheimer's disease. *Lancet Neurol.* 4: 705-711, 2005.

Schuit, A.J., E.J.M. Feskens, L.J. Launer, D. Kromhout. Physical activity and cognitive decline, the role of the apolipoprotein e4 allele. *Med. Sci. Sports Exerc.* 33: 772-777, 2001.

Strittmatter, W.J., A.M. Saunders, D. Schmechel, M. Pericak-Vance, J. Enghild, G.S. Salvesen, A.D. Roses. Apolipoprotein E: High-avidity binding to b-amyloid and increased frequency of type 4 allele in late-onset familial Alzheimer disease. *Proc. Nat. Acad. Sci.* 90: 1977-1981, 1993.

sedentary noncarriers, in effect completely erasing the risk of the ε4 allele (active noncarriers had the lowest risk).

Thus, two of these three studies provide support for a gene × physical activity interaction, such that risk of cognitive decline and Alzheimer's disease in ApoE ε4 carriers is considerably reduced compared to that in sedentary ε4 carriers, who have elevated risk. Watch for many more studies to be published in this area as researchers seek to identify the potential mechanisms for how physical activity and ApoE genotype interact to alter Alzheimer's disease risk.

As a quick side note, an interesting area of related literature is that of ApoE genotype and the response to brain injury, which is relevant to many contact sports. Evidence is emerging that carriers of the ε4 allele have a more adverse initial response to brain trauma (e.g., concussion, traumatic brain injury) compared to noncarriers. Moreover, the recovery from such injury appears to be delayed or incomplete in some ε4-allele carriers. This research is relatively new, and the relevance to specific sport-related injuries has not been conclusively studied, but indications are that ApoE ε4-allele carriers may be at higher risk of future brain-related disorders if they endure head injury, which is relatively common in such sports as boxing, rugby, and American football.

ACE AND SPORT PERFORMANCE

Whether or not elite athletes have a genetic advantage over others has long been a topic of debate for observers of sport, but the general consensus is that some individuals appear to have abilities that are superior to those of their peers, despite similarities in training and other environmental factors. Identifying specific genetic factors that influence sport performance is a very challenging area of research, however, because by its very nature such research is performed in small groups of individuals, each with different sport expertise. For example, can all Olympic athletes be grouped together because of their elite performance status? Perhaps for questions about mental performance under stress, all elite athletes may share common traits; but for most questions, the genetic factors of interest will tend to be sport specific: sprint or power performance, endurance performance, coordination, and so on. Thus, simply defining *performance* as a phenotype represents a considerable challenge for researchers when attempting genetics studies.

Once a phenotype is defined, heritability is estimated and candidate genes can be considered specific to that phenotype, as discussed in chapters 8 and 9. In this example, however, the progression from phenotype definition to heritability estimate to candidate gene selection is not nearly so clear. While many of the phenotypes underlying different types of exercise or sport performance have been shown to be heritable, few studies have been aimed at addressing performance as a phenotype itself. Thus, the assumption is that performance is heritable because it is built upon at least modestly heritable intermediate phenotypes (e.g., muscle strength, $\dot{V}O_2max$).

One candidate gene in question for sport performance codes for the angiotensin-converting enzyme (ACE), an enzyme critical to the renin-angiotensin system (figure 11.4). The renin-angiotensin system is important to the regulation of the circulatory system and blood pressure, in that it converts angiotensinogen to angiotensin II, which is a potent vasoconstrictor. While this information would lead us to consider ACE as a candidate gene for cardiovascular phenotypes and thus endurance performance, skeletal muscle has its own renin-angiotensin system, which may be important for tissue growth and muscle hypertrophy. In this case, we could hypothesize genetic variation in the system to be important for differences in muscle mass phenotypes, making ACE a candidate for sprint or power performance. Thus, researchers recognized the potential importance of the ACE gene for both endurance and sprint or power performance phenotypes, making the specific choice of phenotype less important for this particular candidate gene.

So why is ACE important from a candidate gene perspective? Remarkably, we focus on an intronic insertion/deletion polymorphism in the ACE gene. A 287-nucleotide insertion/deletion polymorphism in intron 16 has been consistently associated with differences in blood and tissue levels of the ACE protein, such that ACE levels are lowest in subjects with the I/I genotype, intermediate in I/D heterozygotes, and highest in the D/D genotype group (I/I < I/D < D/D). Because the ACE protein is important in the production of angiotensin II, which is a potent vasoconstrictor, the D allele has been studied as a risk allele for cardiovascular diseases such as hypertension. Moreover, the ACE enzyme degrades bradykinin, which is a vasodilator, so higher levels of ACE enzyme would be predicted to negatively influence cardiovascular phenotypes through two mechanisms: increasing levels of a vasoconstrictor while decreasing

Renin-angiotensin system

Angiotensinogen
Renin
Angiotensin I
ACE enzyme
Angiotensin II
Bradykinin
Vasoconstriction
Cell growth
Vasodilation
Metabolic influences

Figure 11.4 The renin-angiotensin system. The ACE enzyme converts angiotensin I to angiotensin II, which promotes vasoconstriction and cell growth. Moreover, the ACE enzyme also breaks down bradykinin, which is a vasodilator.

levels of a vasodilator. In contrast, angiotensin II is known to promote cell growth in various tissues, so the D allele's higher ACE levels could be predicted to have an advantage for muscle hypertrophy. Thus, the I/D polymorphism in ACE has been extensively studied for various performance-related phenotypes and is arguably the most frequently studied gene in the area of physical activity and sport.

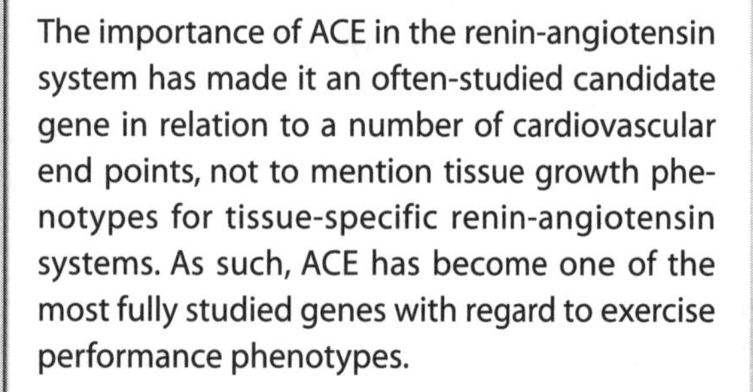

KEY POINT

The importance of ACE in the renin-angiotensin system has made it an often-studied candidate gene in relation to a number of cardiovascular end points, not to mention tissue growth phenotypes for tissue-specific renin-angiotensin systems. As such, ACE has become one of the most fully studied genes with regard to exercise performance phenotypes.

The first indication that the ACE I/D polymorphism may be important for exercise or sport performance came in 1997, when Montgomery and colleagues reported greater left ventricular hypertrophy in military recruits carrying the D/D genotype compared to I-allele carriers after military basic training. In 1998, Montgomery and colleagues followed up with a report in *Nature* showing higher I-allele frequencies in a group of high-altitude mountain climbers compared to controls (a case–control study), as well as greater exercise performance improvements in I/I compared to D/D genotypes after exercise training. This paper provided the impetus for many other groups to test the association of this polymorphism with various performance-related phenotypes in follow-up studies, many of which were reviewed in 2002 by Jones and colleagues. According to Jones' and colleagues' review of many case–control studies, multiple studies showed higher frequencies of the I allele and the I/I genotype in elite-level endurance athletes compared to control populations; multiple studies showed higher frequencies of the D allele and D/D genotype in elite-level sprint or power athletes compared to controls; and several studies showed no association between ACE genotype and either type of sport performance. Examples are presented in figures 11.5 and 11.6.

So what would be driving such allele frequency differences? How is the ACE I/D polymorphism working to result in such differences? Researchers then moved to clarify the phenotype under investigation, most often by examining $\dot{V}O_2$max (maximal oxygen consumption), which is a phenotype often used as a marker of endurance performance. In many of these follow-up studies, ACE genotype was not associated with either baseline $\dot{V}O_2$max or the response of $\dot{V}O_2$max to exercise training, or it was associated with $\dot{V}O_2$max in ways not consistent with the case–control studies (e.g., D allele associated with greater increases in $\dot{V}O_2$max). Certainly, $\dot{V}O_2$max is not the only predictor of performance, and some argue that it is not a good predictor, so other phenotypes were examined, especially targeting skeletal muscle.

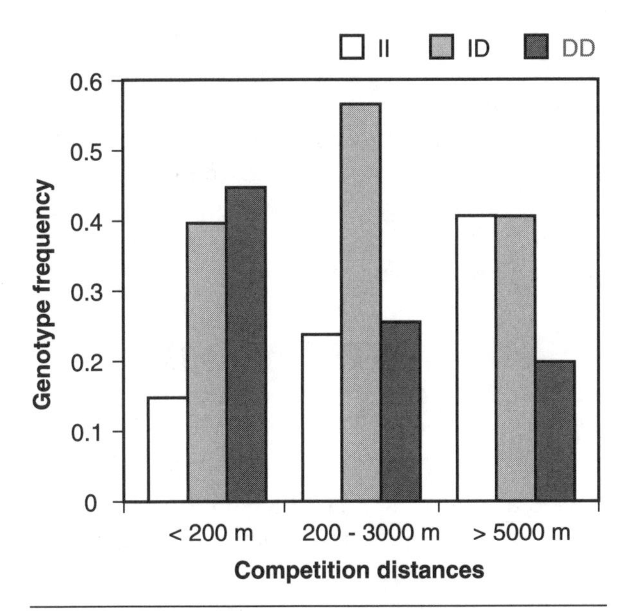

Figure 11.5 Angiotensin-converting enzyme (ACE) I/D polymorphism genotype frequencies in Olympic runners specializing in different distances. Notice the higher I/I genotype frequencies in the runners specializing in distances >5000 meters compared to the shorter-distance groups. In contrast, the short-distance runners (<200 meters) have a higher fraction of D/D genotypes compared to the longer-distance runners.

Reprinted, by permission, from A. Jones, H.E. Montgomery and D.R. Woods, 2002, "Human performance: A role for the ACE genotype?," *Exerc. Sport Sci. Rev.* 30(4): 184-190.

Because skeletal muscle is important for nearly all sport performance and skeletal muscle has a local renin-angiotensin system, skeletal muscle phenotypes important for sport performance may differ among ACE I/D genotype groups. Phenotypes such as muscle efficiency and fiber type proportions have been examined in relation to ACE genotype in an attempt to understand the basis for the associations with elite athletic performance. These studies have shown more consistent results than those aimed at $\dot{V}O_2$max. For example, Williams and colleagues (2000) reported greater muscle efficiency in I/I genotype men compared to D/D men after exercise training. This difference in muscle efficiency, as measured by work performed per unit energy expenditure, was not observed at baseline, suggesting a gene × environment interaction. In fact, only the I/I genotype group showed an increase in muscle efficiency in response to exercise training. In an example of the study of an intermediate phenotype, Zhang and colleagues (2003) reported a higher percentage of type I versus type II muscle fibers in I/I genotype subjects compared to D/D subjects, with intermediate values for I/D heterozygotes. Type I fibers are considered more metabolically efficient than type II muscle fibers, so the higher fraction of type I versus type II fibers in the I/I genotype is consistent with greater muscle efficiency.

Some studies have also specifically targeted the D allele and the hypothesis that it may provide an advantage for muscle hypertrophy in response to exercise. Folland and colleagues (2000) reported a greater increase in muscle strength in D-allele carriers versus the I/I genotype group in men performing a strength training intervention. This result was consistent

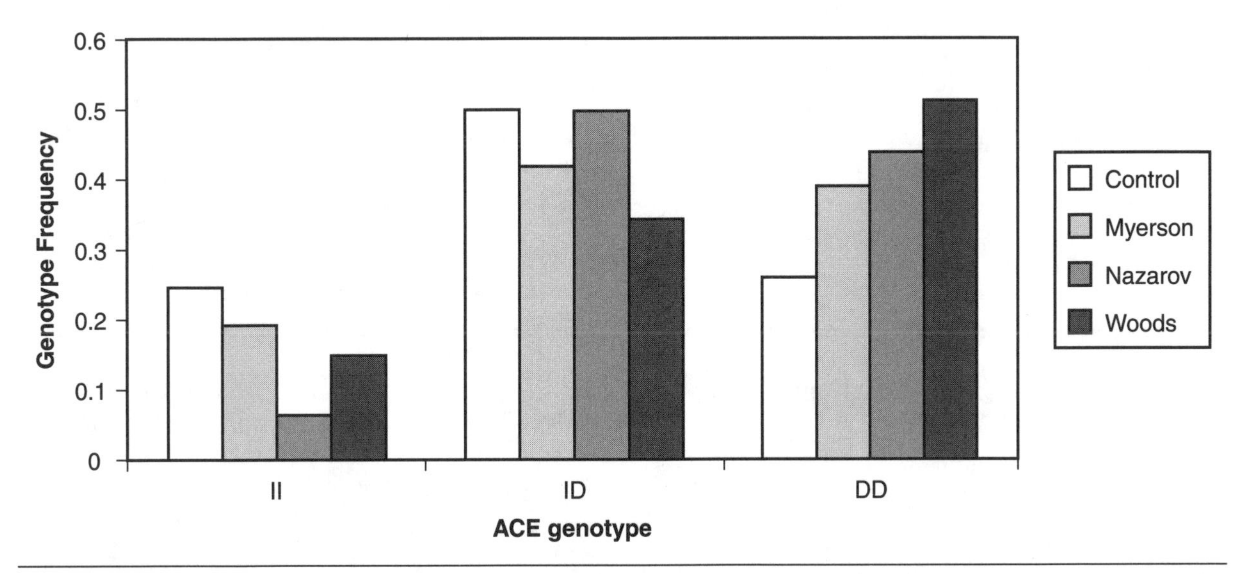

Figure 11.6 Angiotensin-converting enzyme (ACE) I/D polymorphism genotype frequencies for short-distance swimmers as reported in three different research papers in comparison to a control group (white bars). The D/D genotype frequency was higher in the short-distance swimmers compared to the control population for all three research studies (each study shown with a gray or black bar).

Reprinted, by permission, from A. Jones, H.E. Montgomery and D.R. Woods, 2002, "Human performance: A role for the ACE genotype?," *Exerc. Sport Sci. Rev.* 30(4): 184-190.

with expectations that the D allele would provide an advantage because of potential influences on muscle growth. This finding was contradicted, however, by several later studies showing no association between the ACE D allele and enhanced responsiveness of muscle to strength training interventions.

Despite the considerable efforts to study ACE genotype with various sport performance phenotypes, the general consensus has changed very little since that of Jones and colleagues' review of 2002: The I allele appears to be associated with elite endurance performance, potentially through effects on muscle efficiency, and the D allele appears to be associated with elite sprint or power performance, though a mechanism of increased muscle growth has not been consistently shown. As Jones and colleagues wrote (2002, p. 189), "Whatever the data may conclude, elite athletes are still made not born, though perhaps some may be made elite in one discipline more easily than others." In other words, despite the initial excitement about ACE as a gene critical for sport performance, years of research now show that at best, ACE is one of many genes involved in sport performance, and by itself it may have little direct impact on performance outcomes.

MYOSTATIN AND MUSCLE MASS

In 1997, researchers at Johns Hopkins University (McPherron, Lawler, and Lee) reported that mice with a disabled version of a newly identified gene, known as growth and differentiation factor 8, developed

Citations for ACE and Sport Performance

Folland, J., B. Leach, T. Little, K. Hawker, S. Myerson, H. Montgomery, D. Jones. Angiotensin-converting enzyme genotype affects the response of human skeletal muscle to functional overload. *Exp. Physiol.* 85: 575-579, 2000.

Gayagay, G., B. Yu, B. Hambly, T. Boston, A. Hahn, D.S. Celermajer, R.J. Trent. Elite endurance athletes and the ACE I allele—The role of genes in athletic performance. *Hum. Genet.* 103: 48-50, 1998.

Hagberg, J.M., R.E. Ferrell, S.D. McCole, K.R. Wilund, G.E. Moore. $\dot{V}O_2$max is associated with ACE genotype in postmenopausal women. *J. Appl. Physiol.* 85: 1842-1846, 1998.

Jones, A., H.E. Montgomery, D.R. Woods. Human performance: A role for the ACE genotype? *Exerc. Sport Sci. Rev.* 30: 184-190, 2002.

Montgomery, H.E., P. Clarkson, C.M. Dollery, K. Prasad, M.A. Losi, H. Hemingway, D. Statters, M. Jubb, M. Girvain, A. Varnava, M. World, J. Deanfield, P. Talmud, J.R. McEwan, W.J. McKenna, S. Humphries. Association of angiotensin-converting enzyme gene I/D polymorphism with change in left ventricular mass in response to physical training. *Circulation* 96: 741-747, 1997.

Montgomery, H.E., R. Marshall, H. Hemingway, S. Myerson, P. Clarkson, C. Dollery, M. Hayward, D.E. Holliman, M. Jubb, M. World, E.L. Thomas, A.E. Brynes, N. Saeed, M. Barnard, J.D. Bell, K. Prasad, M. Rayson, P.J. Talmud, S.E. Humphries. Human gene for physical performance. *Nature* 393: 221-222, 1998.

Myerson, S., H. Hemingway, R. Budget, J. Martin, S. Humphries, H. Montgomery. Human angiotensin I-converting enzyme gene and endurance performance. *J. Appl. Physiol.* 87: 1313-1316, 1999.

Pescatello, L.S., M.A. Kostek, H. Gordish-Dressman, P.D. Thompson, R.L. Seip, T.B. Price, T.J. Angelopoulos, P.M. Clarkson, P.M. Gordon, N.M. Moyna, P.S. Visich, R.F. Zoeller, J.M. Devaney, E.P. Hoffman. ACE ID genotype and muscle strength and size response to unilateral resistance training. *Med. Sci. Sports Exerc.* 38: 1074-1081, 2006.

Rankinen, T., L. Perusse, J. Gagnon, Y.C. Chagnon, A.S. Leon, J.S. Skinner, J.H. Wilmore, D.C. Rao, C. Bouchard. Angiotensin-converting enzyme ID polymorphism and fitness phenotype in the HERITAGE Family Study. *J. Appl. Physiol.* 88: 1029-1035, 2000.

Rankinen, T., B. Wolfarth, J.A. Simoneau, D. Maier-Lenz, R. Rauramaa, M. Rivera, M.R. Boulay, Y.C. Chagnon, L. Perusse, J. Keul, C. Bouchard. No association between the angiotensin-converting enzyme ID polymorphism and elite endurance athlete status. *J. Appl. Physiol.* 88: 1571-1575, 2000.

Williams, A.G., M.P. Rayson, M. Jubb, M. World, D.R. Woods, M. Hayward, J. Martin, S.E. Humphries, H.E. Montgomery. The ACE gene and muscle performance. *Nature* 403: 614, 2000.

Zhang, B., H. Tanaka, N. Shono, S. Miura, A. Kiyonaga, M. Shindo, K. Saku. The I allele of the angiotensin-converting enzyme gene is associated with an increased percentage of slow-twitch type I fibers in human skeletal muscle. *Clin. Genet.* 63: 139-144, 2003.

much larger muscles than control mice. In fact, the individual muscles of the mutant mice were as much as two to three times larger than those of their normal cousins. All of this muscle growth occurred without other obvious changes in the mice (i.e., the mice were otherwise healthy), so the defect seemed to be specific to the growth of skeletal muscle. Thus, the newly identified protein appeared to be a negative growth regulator of muscle, in that its presence in normal mice prevented the extreme muscle growth found in mice where the gene was disabled; the researchers named the gene and protein that it encodes *myostatin* ("myo," muscle; "statin," something that inhibits).

Quickly following the results in mice, several research groups began to study the genetic sequence of the myostatin gene, especially in cattle. The Belgian Blue cattle breed (see figure 11.7) exhibits considerably higher muscle mass than other breeds, a condition known as "double muscling." While researchers had considered that this unique trait was based on a genetic mutation, the specific gene had not been identified. Several linkage analysis-type studies had been performed, however, showing that a region on the bovine (cow) chromosome 2 appeared to contain a critical gene. The identification of myostatin as a muscle growth regulator in mice provided a specific candidate gene that could be studied in cows. Three groups reported nearly simultaneously that, as predicted, an 11-nucleotide deletion in exon 3 of the bovine myostatin gene, located on chromosome 2, had been found in the Belgian Blue breed, apparently explaining the genetic basis for the double-muscled phenotype. Moreover, other mutations in the myostatin gene are associated with excess muscle mass in other similar cattle breeds.

This extensive work in animal models provided the rationale for examining the human myostatin gene for sequence variation, potentially explaining differences in baseline levels of muscle mass or perhaps the response of muscle to resistance exercise interventions. Muscle mass and muscle strength are both strongly heritable phenotypes, with h^2 values for muscle mass ranging from 50% to 80% and for muscle strength ranging from 30% to 80%. Thus, researchers interested in the genes underlying muscle mass in humans found an attractive candidate gene in myostatin. In the context of study design, myostatin was identified as a candidate gene from animal models showing the importance of the gene to muscle development.

KEY POINT

Myostatin was identified in animal models as a key negative regulator of skeletal muscle growth. Researchers then reasoned that genetic variation within such an important growth regulator could alter either normal muscle growth in humans or the response of muscle to various exercise stimuli.

Figure 11.7 Belgian Blue bull with the bovine myostatin gene mutation.

Reprinted, by permission, from A.C. McPherron and S.J. Lee, 1997, "Double muscling in cattle due to mutations in the myostatin gene," *Proc. Natl. Acad. Sci. USA* 94(23): 12457-12461. Copyright 1997 National Academy of Sciences, U.S.A.

In 1999, Ferrell and colleagues sequenced the myostatin gene in humans and discovered several missense polymorphisms in the exons of the gene, two of which had fairly common rare allele frequencies in the population. The researchers performed a case–control study of elite-level bodybuilders and strength athletes compared to control subjects, but the allele frequencies for the newly identified SNPs did not differ between the two groups, despite the considerable differences in muscle mass.

The identification of the myostatin SNPs then allowed other groups to perform retrospective genetic association studies in other cohorts, including young and older subjects undergoing strength training interventions and older men and women at baseline. Ivey and colleagues (2000) reported that women carrying the rare R allele of the K153R missense SNP in the myostatin gene showed greater increases in muscle mass in response to a strength training program than women with the K/K genotype, but the sample size was very small and the data were considered preliminary. In contrast, Seibert and colleagues (2001) as well as Corsi and colleagues (2002) reported lower muscle strength in older subjects carrying the R allele, though the sample sizes were again small. The rare R-allele frequency is only 4% in Northern European and 12% in African American populations, and most studies to date have focused only on populations of Northern European ancestry. Thus, the consensus in the literature through these studies was that genetic variation in the myostatin gene was not as important in human muscle mass differences as it was in cattle. An important point to consider, however, is that the bovine gene mutation is a nonsense deletion, while the K153R SNP in the human myostatin gene (known as GDF8) is a missense polymorphism. Thus, with these mutations, the cattle myostatin protein is nonfunctioning whereas the human myostatin protein is changed by a single amino acid.

An important new study was added to the literature in 2004 by Schuelke and colleagues. They reported a case study of a healthy German child born with extraordinary muscularity. Figure 11.8 shows images of the child at early ages, including a cross section of the child's thigh compared to that of a matched control child (figure 11.8*c*). The child, followed over several years, showed normal development but with continued high levels of muscle mass and strength. The child's mother is a former professional athlete, and several family members were identified as unusually strong (denoted as black symbols in figure 11.8*d*). Subsequent genetic analysis revealed a rare mutation in intron 1 of the myostatin gene. While intronic mutations do not typically cause major effects on gene function, in this case the mutation falls on a splice site, thus affecting the splicing of exon 1 with exon 2 during RNA posttranscriptional modification. In this case, the splice site mutation results in a nonfunctional myostatin protein. The child is homozygous for this mutation

Citations for Myostatin and Muscle Mass

Corsi, A.M., L. Ferrucci, A. Gozzini, A. Tanini, M.L. Brandi. Myostatin polymorphisms and age-related sarcopenia in the Italian population. *J. Am. Geriatr. Soc.* 50: 1463, 2002.

Ferrell, R.E., V. Conte, E.C. Lawrence, S.M. Roth, J.M. Hagberg, B.F. Hurley. Frequent sequence variation in the human myostatin (GDF8) gene as a marker for analysis of muscle-related phenotypes. *Genomics* 62: 203-207, 1999.

Ivey, F.M., S.M. Roth, R.E. Ferrell, B.L. Tracy, J.T. Lemmer, D.E. Hurlbut, G.F. Martel, E.L. Siegel, J.L. Fozard, E.J. Metter, J.L. Fleg, B.F. Hurley. Effects of age, gender and myostatin genotype on the hypertrophic response to heavy resistance strength training. *J. Gerontol. Med. Sci.* 55A: M641-M648, 2000.

McPherron, A.C., A.M. Lawler, S-J. Lee. Regulation of skeletal muscle mass in mice by a new TGF-b superfamily member. *Nature* 387: 83-90, 1997.

McPherron, A.C., S-J. Lee. Double muscling in cattle due to mutations in the myostatin gene. *Proc. Nat. Acad. Sci. USA* 94: 12457-12461, 1997.

Roth, S.M., S. Walsh. Myostatin: A therapeutic target for skeletal muscle wasting. *Curr. Opin. Clin. Nutr. Metab. Care* 7: 259-263, 2004.

Schuelke, M., K.R. Wagner, L.E. Stolz, C. Hubner, T. Riebel, W. Komen, T. Braun, J.F. Tobin, S-J. Lee. Myostatin mutation associated with gross muscle hypertrophy in a child. *N. Eng. J. Med.* 350: 2682-2688, 2004.

Seibert, M.J., Q-L. Xue, L.P. Fried, J.D. Walston. Polymorphic variation in the human myostatin (GDF-8) gene and association with strength measures in the Women's Health and Aging Study II cohort. *J. Am. Geriatr. Soc.* 49: 1093-1096, 2001.

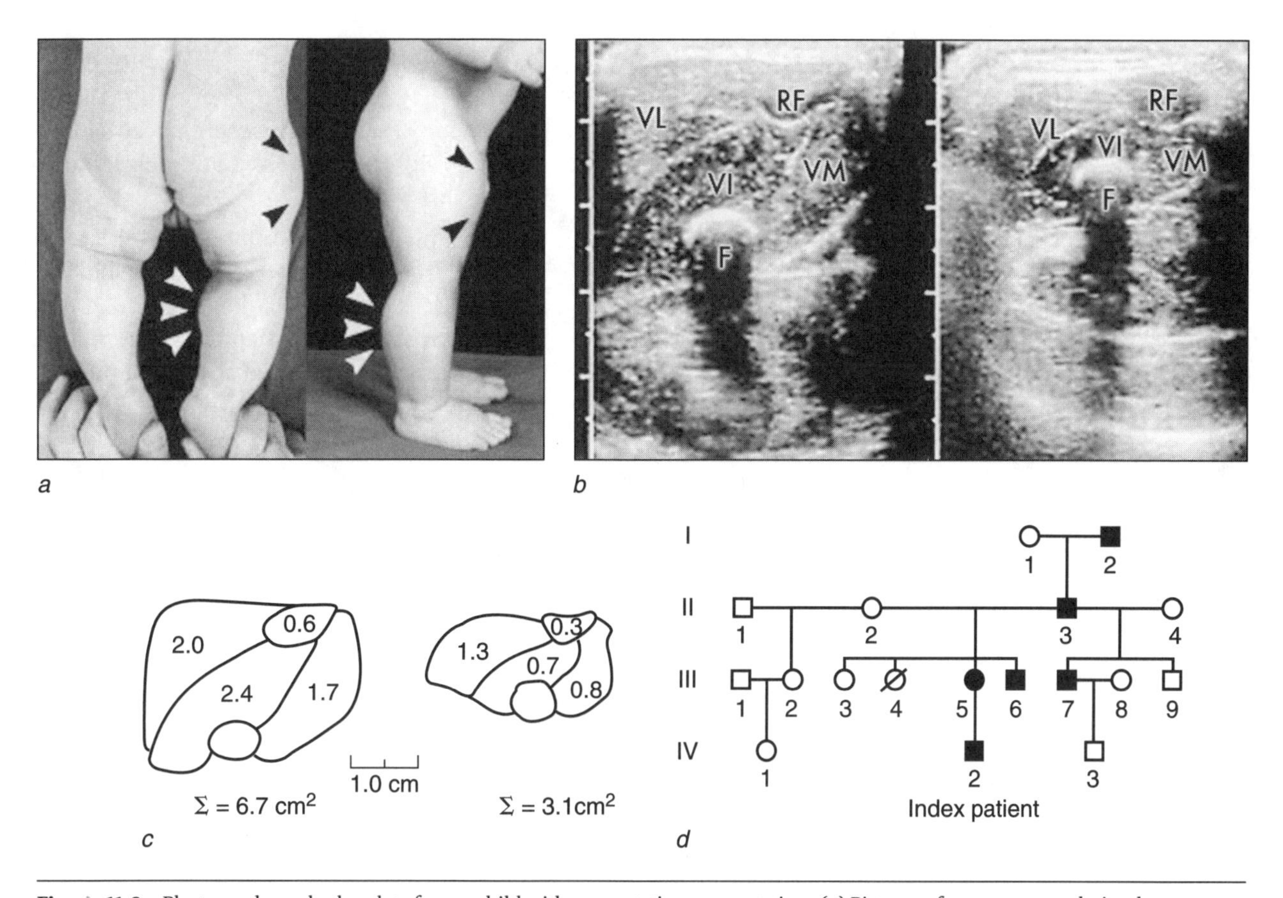

Figure 11.8 Photographs and other data from a child with a myostatin gene mutation. *(a)* Pictures of extreme muscularity shown at two different ages. *(b, c)* Cross-sectional images and drawings of the thigh muscles in the child (left) and a matched control infant; notice the considerably larger muscle cross section in the case child. *(d)* The pedigree of the child's family, with the child labeled as "2, Index patient."

Reprinted, by permission, from M. Schuelke et al., 2004, "Myostatin mutation associated with gross muscle hypertrophy in a child," *N. Engl. J. Med.* 350(26): 2682-2688. Copyright 2004 Massachusetts Medical Society. All rights reserved.

and thus has no functioning myostatin protein (the mother is heterozygous for the mutation; there is no information about the father). The mutation was not found in 200 randomly sampled individuals.

The addition of the work of Schuelke and colleagues to the myostatin literature provides greater support for the idea that myostatin genetic variation will affect human skeletal muscle, but such extreme mutations appear to be very rare in human populations. Unless new polymorphisms are discovered in the myostatin gene, the next likely targets for researchers studying the influence of myostatin-related genetic variation on muscle phenotypes will be the genes that myostatin interacts with in muscle, including its receptor targets, signaling pathways, and any molecular regulators. Presumably, given the importance of myostatin in muscle regulation, genetic variation in these related genes could similarly affect muscle phenotypes.

SUMMARY

In this chapter, we've examined several examples of research findings relevant to the field of genetics in health, physical activity, and sport and also discussed the annual review article that highlights progress in the field. By the time you read this chapter, these findings will necessarily be outdated. The point of the section is to provide some insights into how genetics studies are performed and how the questions and hypotheses evolve as new findings are reported. As we move forward with the remaining chapters of part III, we will examine how genetics research is anticipated to contribute to the future of healthcare, physical activity, and sport, in both positive and potentially negative ways.

REVIEW QUESTIONS

1. What was the rationale for predicting a role for the ACTN3 R577X X allele in sprint performance?
2. Discuss the relationship between ApoE gene variation and Alzheimer's disease risk.
3. Based on the literature reviewed for the ACE I/D polymorphism, how is it possible that the I allele could be associated with endurance performance but not $\dot{V}O_2max$?
4. Has genetic variation in myostatin been associated with differences in skeletal muscle mass in humans?
5. Choose a phenotype of interest to you. Using PubMed, the gene map article described in the Special Focus section (Rankinen et al., 2006) and other methods (see chapter 8), summarize what scientists have learned about that phenotype in recent years.

CHAPTER

12

PERSONALIZED MEDICINE

The incentive to perform the immense amount of work needed to identify genetic variants important to complex traits is that key variants may eventually be used clinically to improve the prevention and treatment of various diseases, thus improving medical care and health across the population. In this chapter, we discuss the use of genetic information in the clinical aspects of health and fitness, providing an overview of what promises to be an ever-growing use of genetics in medicine and health care. Please note that I've used simplified examples in this chapter to illustrate the general principles underlying the use of genetics in health care applications; readers should keep in mind the complexity of genetic and environmental influences on complex phenotypes, as discussed in chapter 7.

GENETICS AND COMPLEX DISEASE REVISITED

In the early chapters, I presented the idea that simple DNA sequence variation within a gene region could have impacts on the regulation of that gene, or its coded protein, or both, the end result being differential effects on the phenotype depending on which allele was present in the polymorphism. That concept provided the foundation for understanding the potential importance of genetic factors in all heritable phenotypes, and the discussion of Mendelian diseases solidified that idea by showing the sometimes devastating consequences of even a single base mutation in the DNA sequence.

In addition to gaining insight into the basis for rare single-gene diseases, researchers are interested in understanding the etiology or basis for complex disease phenotypes, as these affect the greatest numbers of people in the population—including such diseases as cardiovascular disease, obesity, and type 2 diabetes. These are heritable yet complex traits, which we know include the combinations of influential genetic and environmental factors.

Recall the pie chart used in chapter 5 to represent the combined contributions of these two types of factors, with each major area made up of many small slices, each representing the individual components and the extent of their influence on the phenotype. The point was that despite the large number of factors influential for a trait of interest, many factors will have only small influences, and a few are likely to have larger influences. The genetic and environmental factors with the largest contributions to a trait are those most likely to be important for a large number of individuals. We can think of these factors as those with the greatest likelihood of being clinically significant or having relevance to public health. For this reason, genetic association studies and even general epidemiology studies aim to identify the most important factors and their interactions, with the goal of applying this understanding to health care for the general public.

Once smoking was shown to increase cancer risk, that environmental factor was targeted within many health care communities to reduce cancer risk and improve overall health. Similarly, once physical activity was shown to affect cardiovascular disease risk, obesity, and mortality, regular physical activity became a common recommendation in health care. Now, as we begin to identify the genetic factors that contribute to susceptibility for complex disease, these factors have the potential to enter the mainstream of health care interventions. The idea of using genetic information to specifically tailor health care interventions is known as personalized medicine.

PROMISE AND LIMITATIONS OF PERSONALIZED MEDICINE

Personalized medicine, also known as *individualized* or *genetic medicine,* has been described as medicine taken to an individual level through the incorporation of an individual's specific genetic variation (i.e., **"genetic profile"**) into his or her clinical treatment. In the extreme, unrealistic version of this scenario, you could be prescribed a drug that was specifically formulated just for you, based on your unique combination of genetic factors and environmental influences. Certainly, health care is already considered "personalized" in that some combination of therapies or interventions is applied to improve health in any individual, but the tools used today are fairly general and are applied in varying combinations to virtually all people seeking help. The difference with personalized medicine is that the genetic information of each patient will be uniquely incorporated into his or her health care, leading to specific drug prescriptions or other interventions that are known to work differently in individuals with different genetic profiles.

The promise of adding genetic information to disease prevention and treatment is that, if a variety of gene- and environment-based interventions are individually tailored to a particular patient, more and more people will be successfully treated for disease, thus improving health across the entire population. Not all individuals respond equally to the major treatment options for any of the complex diseases, and this discrepant response certainly has a genetic basis in some cases. If these unique genetic factors (and therefore more of the total factors contributing to the disease) are accounted for, more people will enjoy successful treatment with improved overall health. As discussed in the Special Focus section, the most work in this area has been in the field of prescription drug development and dosage ("pharmacogenomics"), various drugs already having been shown to have differential effects due to genotype. The preliminary findings are promising and provide examples of what personalized medicine might look like in the future. But the state of personalized medicine today is more about the promise of future developments than about current usefulness in the clinic, and there are some arguments for why genetic factors may not become mainstream in future health care.

KEY POINT

Personalized medicine is the inclusion of an individual's genetic variation information in the design and implementation of health care interventions to improve the efficacy and reduce the risk of the intervention.

From a scientific standpoint, the rationale underlying personalized medicine is well accepted, and certainly many examples of genetic influence on phenotype variability and even disease can be presented to support the concept. That said, success in finding key genes for complex traits underlying common diseases, such as obesity, type 2 diabetes, and cardiovascular disease, has been limited to date owing to the incredible number of genetic and environmental factors contributing to each disease. Perhaps the emerging tools of statistical genetics and high-throughput genotyping will allow more genes to be conclusively identified, but a counterargument would be that the number of genetic and environmental factors important for a particular disease phenotype represents so many unique combinations (i.e., interactions) that finding common disease pathways for large fractions of the population will remain difficult. If only a small fraction of the population shares a particular gene or environment disease susceptibility, the chances for a targeted intervention important only for that group will be diminished simply from an economic standpoint. At minimum, however, the work being performed will certainly add to our understanding of the physiology of disease in general, which can only help to improve health care, whether or not genetic information becomes firmly integrated into everyday health care.

On the business side of personalized medicine, how companies can uniquely benefit from adding genetic factors to an intervention is not always clear. In some cases, the addition of genetic information would limit the usefulness of (and thus the market for) an intervention to individuals with specific genetic profiles, which could limit company earnings by decreasing the number of available consumers. And if patients, or consumers, must provide genetic information in order to qualify for a gene-based intervention, some will no doubt decide not to provide that information out of fear of genetic discrimination in other aspects of their lives.

It is important to note that while personalized or genetic medicine is not currently found in mainstream health care, advances in genetic technology will likely change that. One idea is that an individual's entire genomic sequence will become readily available for low cost. In fact, the J. Craig Venter Science Foundation has announced a $500,000 prize for the development of a "$1000 genome." The goal is the development of technologies that will allow sequencing of an individual's entire genome sequence for as little as $1000, making personalized genetic information more readily available for mainstream medical care. The U.S. National Institutes of Health has also announced grant opportunities dedicated to this same goal. Companies from around the world are testing such technologies with the goal of accomplishing the $1000 genome by 2010.

The remaining sections of the chapter delve into some of the expected applications of personalized medicine in various aspects of health care, from treatment to prevention of complex disease.

SPECIAL FOCUS
Pharmacogenomics

Traditionally, prescription drugs have been used for disease treatment without regard to genetic differences among individuals, but that is changing, as both drug side effects and drug effectiveness have been associated with genetic variation. Thus, more and more pharmaceutical companies around the world are now studying gene × drug interactions, an area known as **pharmacogenomics.** Nearly all drugs have side effects, which are unintended, and in some cases harmful, effects on the body unrelated to the phenotype targeted by the drug, but those side effects are typically seen in only a fraction of patients. Moreover, a remarkable number of people show only limited responsiveness to many common prescription drugs. Thus, there is considerable variation in drug responsiveness, in terms of both effectiveness and side effects. Through accounting for individual genetic variation, the goal of pharmacogenomics is to improve drug effectiveness and reduce side effects by targeting drug prescriptions and dosages specific to an individual's genetic profile.

An excellent example of the promise of pharmacogenomics centers on a drug known as 6-mercaptourine (6-MP), which is used to treat a form of childhood leukemia. A mutation in the gene TPMT can eliminate the function of the enzyme needed to metabolize 6-MP, such that the mutation can result in life-threatening levels of the drug in individuals homozygous for the mutation. Fortunately, only a small fraction of individuals (<1%) carry two copies of the mutant allele, so the general risks are low, but the consequences can be drastic for the homozygote carriers. A gene test was developed for TPMT in the mid-1990s and, despite early resistance by physicians, is now being used more frequently to screen leukemia patients before the initiation of 6-MP therapy. Individuals screened as homozygotes for the mutation can be prescribed alternate therapies, which will protect them from the dangerous side effects of 6-MP.

Another example is the use of the blood thinner (anticoagulant), warfarin, in the treatment of cardiovascular disease and prevention of stroke. Dosages of warfarin that are too low are ineffective, yet doses too high can lead to severe bleeding complications, and finding the correct dose for any one person can be challenging, requiring repeated blood tests and prescription changes over a lifetime. Sequence variation in two genes, CYP2C9 and VKORC1, has been shown to account for differences in warfarin metabolism (and thus required dosages) among people, thus making available specific genetic information that could be used prior to the initial prescription of warfarin and therefore improving effectiveness and reducing risk. While genetic screening prior to drug prescription is not yet common, as more examples such as these become well established, patients will benefit from the option of more individualized prescriptions.

Geographic ancestry is also being considered in pharmacogenomics. Drug metabolism has been shown to be lower in some East Asian populations, apparently owing to higher frequencies of certain alleles present in the genes of various drug-metabolizing enzymes. Specifically, the activity of certain cytochrome P450 enzymes, important for the liver's clearance of many commonly prescribed drugs from the body, has been shown to be lower in East Asians compared to Northern European populations owing to differences in allele frequencies between

(continued)

SPECIAL FOCUS

(continued)

the groups. For example, East Asians have a higher frequency of the low-metabolizing haplotype in the VKORC1 gene than other populations, resulting in a lower safe and effective dose of warfarin for that population.

Similarly, the prescription drug BiDil was approved in 2005 for prescription in the United States specifically for the treatment of heart failure in African Americans—the first such population-specific drug approval in the United States. The justification for the BiDil decision was that a large clinical trial of African Americans showed remarkable reduction in mortality risk for subjects taking BiDil, demonstrating the drug's effectiveness in this specific population. The drug's mechanism of action is to increase nitric oxide levels; low levels of nitric oxide are often cited as an important factor in heart disease in African Americans, potentially due to genetic factors. Whether the drug would show similar benefits for other populations has not been conclusively studied; this raises questions about whether the race-specific recommendation was appropriate. In the case of BiDil, race is used as a proxy for geographic ancestry, which is correlated to genetic variation more commonly found in individuals of a specific ancestral population. Ultimately, knowing the specific susceptibility or risk alleles underlying the risk of heart failure will allow for individualized prescription, rather than prescription by race or geographic ancestry (as is the case for VKORC1 and warfarin).

GENETICS IN NUTRITION AND EXERCISE PRESCRIPTION

Although the targeted prescription of medications in health care is the example of personalized medicine most commonly thought of, other treatment interventions are also typical in the clinic and can also be considered for prescription based on genetic factors. Two of the most common therapeutic interventions are the environmental factors of diet and exercise. In fact, in the case of many complex diseases such as hypertension, type 2 diabetes, and obesity, diet and exercise are often considered first-line interventions. If these lifestyle interventions are unsuccessful by themselves, medications are typically prescribed to the patient with the condition that the lifestyle interventions continue while the person is on the drug therapy. Thus, just as drug prescriptions can be targeted to individual genetic profiles to improve efficacy and limit side effects, diet and exercise prescriptions can be targeted to maximize effectiveness in individual patients. The application of individualized dietary prescription has begun to enter the fringes of health care practice, though the application of individualized exercise prescription is still in its infancy. Let's examine both areas in more detail.

Nutrigenomics

The application of genetic information to dietary intake is known as **nutrigenomics** or nutritional genomics, from the parent word, "nutrition." The basic premise behind nutrigenomics is that the components of dietary intake become environmental factors, each with the potential to interact with various target genes. If those target genes have sequence variation, then the possibility of differential effects among different alleles exists, with impacts on the related phenotypes. Recall from chapter 6 the discussion of phenylketonuria (PKU), in which a genetic mutation in the gene that encodes the phenylalanine hydroxylase enzyme (PAH) prevents metabolism of the amino acid, phenylalanine, with consequences for cognitive development if left untreated. This is an example of a dietary component interacting with a genetic sequence variant with consequences on a phenotype.

Phenylketonuria is an extreme, Mendelian disease example of how genetic variation can interact with dietary factors to affect a phenotype, but the same logic holds for complex phenotypes, and several examples have been reported in the research literature. Perhaps the most widely cited example is that of a C/T missense polymorphism in the MTHFR gene, which codes for an enzyme important in the conversion of homocysteine to methionine. High blood levels of homocysteine are considered an independent risk factor for cardiovascular disease, and the T allele in the MTHFR gene is associated with lower enzyme activity, which results in higher homocysteine levels. A key second piece to this puzzle, however, is that folic acid, a dietary component, is necessary for the conversion of homocysteine to methionine (i.e., it is a cofactor for the conversion reaction). Thus, only T allele-carrying individuals with *low* folic acid intake are consistently shown to have high homocysteine levels. The T allele is carried by as many as 15% of individuals with Northern European ancestry. This is an example of a nutrient × gene interaction, in which the combination of the T allele and low dietary intake of folic acid results in elevated levels of a cardiovascular disease risk factor.

The MTHFR T allele would appear to be an excellent candidate for use in personalized medicine and nutrition prescription. We can hypothesize that through an increase in folic acid intake in T-allele carriers, homocysteine levels will be reduced, thereby decreasing cardiovascular disease risk. In fact, companies are already marketing genetic tests including this MTHFR polymorphism specifically for folic acid dietary recommendations. The problem is that the research needed to formally test the hypothesis is incomplete and inconclusive. There is no strong evidence that treating people with folic acid prevents cardiovascular disease, and not all studies have shown that T-allele carriers are at greater risk than others for the disease.

Another example in nutrigenomics is the association between genetic variation in the apolipoprotein E (ApoE) gene and blood cholesterol levels. The ApoE protein is a component of very low-density lipoproteins (VLDL) and plays an important role in lipoprotein metabolism. Two nearby missense single-nucleotide polymorphisms (SNPs) in the ApoE gene result in three common alleles present in humans, known as ϵ2, ϵ3, and ϵ4. The ApoE ϵ4 allele has been associated with increased cardiovascular disease risk (as well as elevated Alzheimer's disease risk, as seen in chapter 11). ApoE ϵ4-allele carriers tend to show higher levels of LDL cholesterol (i.e., "bad" cholesterol) compared to either ϵ2 or ϵ3 carriers. As with the MTHFR example, however, the relationship between ApoE genotype and LDL levels appears to be modified by dietary intake, such that ϵ4 carriers show reductions in total and LDL cholesterol levels in response to a low-fat diet to a greater extent than ϵ2 or ϵ3 carriers. Other data suggest that the cholesterol levels of ϵ2 and ϵ3 carriers are more responsive to changes in the type of dietary fat (e.g., saturated, unsaturated) than those of ϵ4 carriers. But not all studies show the same associations, which makes sense given the many factors (gene and environment) likely to influence cholesterol levels. Thus, dietary patterns appear to influence the relationships between ApoE genetic variation and blood cholesterol levels, though research continues in order to clarify inconsistencies found among studies.

Other gene × diet associations have been reported, and, as seen from the preceding examples, there is considerable promise for the future use of genetic information in individualized dietary interventions designed to reduce disease risk. Most importantly, researchers continue to strengthen the science underlying these preliminary findings, thus moving toward greater consensus on the usefulness of such gene-based dietary interventions in future disease treatment and prevention strategies.

Kinesiogenomics

The second area of focus for this section is genetics in exercise or physical activity prescription. This research area is known as **kinesiogenomics,** from the parent word, "kinesiology," meaning the study of human movement. This area is substantially less developed than nutrigenomics, with fewer groups and companies worldwide focused on the primary research questions.

Recall some of the data presented in chapter 1, where the considerable variability in responsiveness to exercise training was shown for various phenotypes. Some individuals showed dramatic improvements in the phenotype, while others showed little or no change in the phenotype. Several studies have shown that the responsiveness of various phenotypes to exercise training are heritable, suggesting the importance of genetic factors, and the identification of those contributing factors is an area of ongoing investigation in the field. Thus, individuals with specific genetic profiles are expected to be very responsive to a particular exercise intervention for specific phenotypes, and perhaps nonresponsive for other phenotypes.

For example, a person might be remarkably responsive for change in $\dot{V}O_2$max with aerobic exercise training while showing very little response of blood pressure values to that same training intervention. If this person was headed into a physician's office for treatment of hypertension, knowing about this nonresponsiveness to exercise training for blood pressure phenotypes (via genetic screening) would likely alter the treatment interventions prescribed for the individual. Certainly the benefits of exercise would be beyond blood pressure phenotypes, so the patient would still be encouraged to exercise for other benefits, but for the treatment of hypertension, exercise training would not be an effective intervention by itself. In this example, it would be possible to prescribe medications or other therapies sooner for this individual rather than waiting the typical three to six months to evaluate the effectiveness of the exercise intervention on blood pressure.

The opposite could be the case as well: Knowing that an individual is highly responsive to exercise for blood pressure phenotypes might allow full treatment of hypertension without the need for prescription drugs, eliminating this source of side effects and cost in this person's health care management. Thus, knowing genetic factors that predict responsiveness to exercise training can make medical prescriptions more efficient and effective.

The expected applications of genetics to exercise prescription are typically centered on one of two areas: (1) targeting exercise interventions to improve disease risk factors, as just discussed, or 2) targeting exercise interventions to maximize sport performance. There is no published use of individualized exercise prescription in medical care, partly because the key genes and alleles known to be important for the various disease risk factors are still being identified and validated. The current state of the science in kinesiogenomics is reviewed regularly in the journal *Medicine and Science in Sports and Exercise,* as discussed in chapter 11. This review is a good way of identifying key genetic associations with exercise-related phenotypes that may be used clinically in the future.

The second, more controversial, area of genetic intervention in exercise is that related to sport performance. The ethical issues underlying this area are presented in chapter 13. The idea here is that specific genetic factors will predispose certain individuals to excellence within certain sport activities, thus opening the opportunity for genetic screening in the selection of athletes early in the sport recruitment and training process. Companies around the world are now marketing gene testing products with the intent of targeting individuals for specific types of sport performance (e.g., the ACTN3 R577X polymorphism for endurance vs. strength or power events), and some professional sport organizations are using the tests in an attempt to maximize training outcomes and individual performance. In other words, athletes are being screened for genes that have been related to performance, and their training programs or position assignments (e.g., in team sports) are being adjusted to best match their genetic profile (for example, emphasizing endurance rather than sprint or power training for ACTN3 X/X homozygotes). Whether these tests are based on sound science is questionable, especially given the infancy of the research and the lack of clear, conclusive results for many of the genes; but athletes and coaches are not always known for restraint when trying to maximize sport performance. Certainly, this will become an area of intense discussion in the near future as more and more key genes are targeted and their roles in exercise responses and sport performance are more clearly defined. We will discuss these issues in greater detail in the next chapter.

KEY POINT

Incorporating an individual's specific genetic information into diet or exercise prescription may increase the efficacy of these interventions to prevent or treat disease. These areas are known as nutrigenomics and kinesiogenomics.

GENETICS IN DISEASE PREVENTION

Disease prevention should be an important goal of health care research, with treatment of disease becoming less necessary as fewer people succumb to disease owing to increased awareness and use of prevention strategies. Thus, personalized medicine also encompasses the idea of disease prevention, because understanding the various factors underlying a disease (both genetic and environmental) will provide information about the prevention of that disease.

This topic is afforded its own section in the chapter because it implies a critical component not seen in the individualized treatment of disease: genetic screening prior to disease onset. In other words, if genetic factors increase susceptibility to a disease in an individual, then in order to prevent disease in that individual, genetic screening will need to be performed early in life in order for appropriate individualized prevention interventions to be prescribed. The genetic information has to be known first; otherwise genetic susceptibility cannot be known (aside from that associated with typical family medical history). This differs from the situation with use of genetic information to treat disease, in that when an individual presents to a physician or other health care worker with disease symptoms, treatment is required to improve health and maintain quality of life. Few options may exist for disease treatment, so someone otherwise reluctant to submit to genetic testing may do so knowing that the best treatment options may be impossible without specific genetic information.

Whether someone will submit to genetic testing *before* the onset of disease symptoms is an open question. Certainly, some people will and some people won't, but for prevention strategies to be useful, presumably a large number of individuals will need to participate. Will a majority opt for genetic screening for disease prevention? And for what diseases will people willingly undergo testing? What prevention strategies will be available based on genetic information? The available prevention strategies for a particular disease will likely influence a person's decision, as the thought of a lifetime of individualized drug, nutrition, or exercise therapy, even as a preventive measure, may not be acceptable to some individuals. As outlined in the next chapter, the use of genetic information in medical care is a topic of heated debate, and various ethical implications exist that make the decision regarding genetic screening difficult for many people.

Implicit in this discussion of personalized disease prevention, however, is the idea that individualized prevention strategies will need to be studied and verified before genetic screening can even be implemented in a population, let alone be effective. Moreover, the genetic factors important for disease susceptibility or treatment may not be the same genetic factors that will need to be tested for ensuring disease prevention, as the prevention interventions are likely to be different from treatment interventions and thus different physiological pathways may be affected. In other words, knowing that a person is susceptible to disease is just one part of the puzzle. The second part is determining the most effective prevention strategies for that person, given his or her unique genetic factors and environmental influences. So, personalized medicine for the treatment of disease will enter mainstream health care far sooner than will personalized disease prevention.

KEY POINT

The use of genetics for disease prevention differs from its use in disease treatment, in that genetic screening before disease onset is a requirement for use in disease prevention.

SUMMARY

The application of an individual's genetic information to health care interventions is known as personalized medicine. Although few examples of successful personalized medicine interventions can be shown, the promise of this individualized approach to health care is a driving force behind much of the genetic association research being performed today. Whether the incorporation of genetic factors into medical care will ever reach a large fraction of the world's population remains to be seen. The use of genetic information in the design and prescription of pharmaceuticals, as well as in diet and exercise interventions, was also discussed in this chapter, along with specific examples of the anticipated use of such genetic information in health care. Finally, the idea of using genetics in the area of disease prevention was presented, though the future use of genetic information for prevention strategies is less likely than that for medical treatment.

KEY TERMS

genetic profile
kinesiogenomics
nutrigenomics
personalized medicine
pharmacogenomics

REVIEW QUESTIONS

1. What is meant by the term *personalized medicine*?
2. What is pharmacogenomics? What is an example of a gene × drug interaction that might be targeted for gene screening of individuals before drug prescription?
3. Knowing how diet and physical activity might act as environmental factors for a phenotype, how can they potentially interact with genetic factors, which might have relevance for personalized diet and exercise prescription?
4. For what reasons are personalized disease prevention strategies likely to find their way to the clinic later than personalized disease treatment strategies?

CHAPTER

13

ETHICAL CHALLENGES IN GENETICS AND SOCIETY

Chapter 13 moves into the challenging area of ethics, for which genetic technology raises many questions. The chapter is divided into three main parts, dealing with (1) the ethical dilemmas raised by the expanding use of genetics in health care and thus society in general, (2) the use of gene therapy in medicine, and (3) the use of genetic technologies in sport. In the previous chapter, we saw that one of the limitations of personalized medicine is the question of whether people will actually move forward with genetic testing for disease prevention and treatment. What are the concerns that might prevent someone from participating in genetic testing that is designed to improve health? The major issues in this area are outlined in the first part of the chapter.

The second and third parts of the chapter push the envelope by considering the more extreme forms of genetic medicine, whereby disease is treated by intervention on the genome sequence itself. This has potentially drastic implications not only for health care and sport, but also for what it means to be human. What limits should be put on the use of genetic technologies? And, if these technologies are limited to "disease" treatment or prevention, what are the definitions of a disease? Here, we will also discuss the use and potential abuse of genetic technologies in sport. The chapter should provide insight into many of the challenging ethical issues surrounding genetics, with the goal of giving readers the information needed to interpret new findings and form their own conclusions.

ETHICAL, LEGAL, AND SOCIAL IMPLICATIONS OF GENOMICS

At the very beginnings of the Human Genome Project, the scientists involved expected that the complete sequencing of our genetic information would have more than scientific consequences. Opening up the DNA instruction book would have consequences for disease prediction, genetic medicine, and society in general—consequences that were impossible to anticipate at the time the project was conceived. Given this, a commitment was made from the start to dedicate a portion of the project's budget to research on the **ethical, legal, and social implications (ELSI)** of knowing the sequence of the human genome. This area is known as the ELSI Research Program, and historically 3% to 5% of the overall project's budget has gone to such research since the program was implemented in 1990. More information can be found at the National Human Genome Research Institute's (NHGRI) Web site, www.genome.gov.

ELSI research funds are dedicated to understanding various issues, including intellectual property and genetic information, the use of genetic information in non-health care settings, the impact of genomics on concepts of race and ethnicity, and the ethical boundaries of various groups for the uses of genomics in society. Discussing all of the various topics that fall under the ELSI umbrella is well beyond the scope of this text. Rather, we will discuss some of the major areas of concern within the ELSI research umbrella.

DNA, Privacy, and Genetic Discrimination

Some of the greatest concerns regarding the use of genetics in medicine and health care surround the possibility that genetic information might be used for discriminatory purposes. DNA is unique from other types of medical information in that an individual's genetic sequence is unchanging over time, can be predictive of disease and potentially behavior, and is shared among family members. This is unlike a typical blood pressure measurement or a blood lipid profile: While these latter are important bits of medical information, they represent conditions at one point in time that can be changed by various means and may or may not be shared with other family members. Instead, a DNA sample taken at even a very early age in life can be accessed numerous times to gain information about a person, perhaps even without his or her knowledge (in contrast to the situation for a blood pressure or other similar medical test, which requires the person's presence).

Thus, many individuals are concerned that genetic information may be used to harm them in some way because of their unique genetic profile and potential disease susceptibilities, either through loss of privacy or lack of access to employment or health care. For example, genetic data could be misused by employers or insurance companies to prevent the employment or insurance coverage of individuals at risk for genetic disease. Insurance companies might benefit from such screening by identifying clients with a strong potential for higher medical costs, and may thus demand higher insurance premiums. Similarly, employers, who often contribute to the costs of insurance, might stand to gain from identifying employees likely to cost more in terms of benefits. Cases of **genetic discrimination** have been reported in Australia, Germany, and the United States and include a woman denied a teaching position in Germany because of a family history of Huntington's disease.

> **KEY POINT**
>
> Though rare, cases of genetic discrimination in employment and insurance access have been reported and are a primary cause of concern for the public as it considers the expanded use of genetic information in health care. These concerns have implications for participation in everything from genetic testing for disease susceptibility to genetics research.

Another example is simply the privacy of genetic information. How can medical records containing genetic information be secured, and should genetic sequence information be treated differently from typical medical information? Considerable protections are in place to ensure privacy of medical records, but are these laws comprehensive enough to cover the details of genetic information and DNA samples? An underlying consideration relating to this question is that DNA sequence is predictably shared among family members. Thus, genetic sequence information obtained in one sibling likely has consequences for other siblings, similar to family history information in general. In the case of genetic information, however, the results are specific and potentially more meaningful to other siblings than simple family history. Should patients with a diagnosed genetic susceptibility be required to inform siblings or offspring of such a diagnosis? Does genetic information fall outside the typical boundaries of the patient–physician relationship, such that physicians may have a requirement to inform family members (as is the case for some infectious diseases and threats of violence)? What harm will come to family members if such information is not disclosed? Many of these questions are still open for debate as legislatures and policy advocates continue to contemplate the possibilities of genetic medicine now and into the future.

The widespread concern over genetic discrimination leads naturally to a discussion of the laws banning such discrimination. Some countries have clearly banned genetic discrimination (e.g., Austria, Belgium), and the Council of Europe has established a Convention on Human Rights and Biomedicine that explicitly prohibits genetic discrimination of any kind, though some countries have not agreed to sign the convention (e.g., Germany, Ireland, United Kingdom). The United Kingdom has established a voluntary agreement with industry to ban genetic discrimination through 2011. In the United States, several states have banned the improper use of genetic information in insurance or employment practices, but federal legislation has not been enacted despite bipartisan political support and repeated attempts by the U.S. Congress to pass such legislation. Updates on United States legislation at the federal and state levels can be found online at the NHGRI Policy and Legislation Database (www.genome.gov/PolicyEthics/LegDatabase/pubsearch.cfm)

Ethical Concerns in Genetics Research

From a researcher's point of view, these concerns are relevant to genetics research studies involving human subjects. Will research participants willingly participate in a study in which genetic information will be obtained, even if that information is completely confidential and used only for research purposes? How can participants be assured that their genetic information will remain private? Is it appropriate to return genetic information to subjects (as is commonly done for other phenotype data) without formal genetic counseling on the limitations of genetic information for disease prediction? These are some of the key concerns that researchers must be aware of when embarking on a new genetics research study.

In a genetics research study, participants are providing a DNA sample that will be screened for specific genetic information (genotypes), which will then be studied in relation to some health or exercise-related phenotype. All such research, whether it includes a genetic component or not, must be approved by the institutional review boards (IRBs) at the institutions where the work will be performed, demonstrating compliance with all legal requirements for the performance of such work. Within genetics research, the same concerns mentioned earlier about genetic privacy and the potential for genetic discrimination hold true for participation in research studies. Approvals from IRBs for human genetics studies are thus dependent on information from researchers showing how they will protect each individual's privacy and secure the DNA samples and gene sequence data. Researchers are required to label DNA samples obtained from research participants in such a way that they are completely confidential and cannot be identified by anyone outside of the research team. Moreover, some IRBs require that DNA samples not be stored indefinitely, but rather be destroyed at some point after the research project has ended. Discussing these issues during the informed consent visit is often critical to reassuring subjects that protections are in place to secure their DNA and genetic information.

Another concern for researchers is whether or not to provide genetic information to the research participants, who will often want that information as a part of their participation. Very often, genotypes obtained for research are not identified using clinically approved laboratory procedures, as required for most clinical genetic testing, so most scientists argue that the genetic data cannot be provided directly to research participants. In other words, laboratories used in medical clinics must apply for specific certifications that verify the accuracy of their laboratory methods—something that research laboratories rarely undergo. Moreover, research participants will rarely understand the limitations of genetic information as a predictor of disease risk and so on, which places an exceptional burden on researchers to potentially act as genetic counselors, something they are not trained to do. Thus most researchers refuse to provide genetic information to their participants on the grounds that the information is not from a clinical laboratory, is experimental in nature (i.e., not conclusive), and would be provided without the genetic counseling required to fully inform participants of the meaning or usefulness of the genotype data. In fact, policy recommendations have been made to this effect, because full "care" for the participants is not in place when these limitations exist, and care for participants is the primary obligation of the IRB process. This will undoubtedly be a source of frustration for some participants expecting to receive the genetic information. Explaining the limitations of genetic data obtained during experimental research should be a part of the informed consent process and will hopefully alleviate the concerns of most participants.

Seeking Out Genetic Information

Assuming that protections are provided for all of the areas of concern regarding genetic privacy and discrimination outlined so far, some question whether or not people will actively seek out genetic testing for information about disease susceptibility. Will people use genetic information to alter health behaviors or seek therapeutic interventions for genetic susceptibilities? What are the emotional consequences of discovering a genetic disease susceptibility, especially if the disease is severe or untreatable? Knowing you are susceptible to baldness is different from knowing you are susceptible to Alzheimer's or Parkinson's disease. Would individuals thus seek genetic testing for only some possible genetic susceptibilities? Proponents of personalized medicine could argue that knowing information about all genetic susceptibilities would allow individuals to seek all possible preventive therapies in advance of disease symptoms, thus slowing or preventing disease onset. On the other hand, for severe diseases with limited treatment potential, the emotional burden of knowing about a genetic susceptibility may be traumatic in itself, which could lead to other health problems (e.g., anxiety, depression). In the context of physical activity and diet, would all of these issues regarding DNA privacy prevent people from undergoing genetic testing that might help them optimize general health and wellness (i.e., disease prevention) through individualized exercise and diet prescriptions? Such targeted genetic testing could be envisioned without testing for specific disease susceptibility, but defining health independent of disease susceptibility is awkward and certainly doesn't fully exploit the advantages of genetic information. These are challenging areas that will require considerable public conversation and input before policies can be put in place on a broad scale.

The information presented thus far is certainly not a comprehensive analysis of the ethical issues surrounding the use of genetic information in health care but rather represents the major concerns of individuals and policy makers across the broad spectrum encompassed by genetics in society. From here, we move to the second part of the chapter, which focuses on the issues surrounding the use of genetic therapy for disease treatment.

GENE THERAPY FOR DISEASE TREATMENT

For many genetic diseases, especially Mendelian-type single-gene disorders, the cure can easily be envisioned: Fix the inborn genetic mutation (e.g., nucleotide or insertion/deletion), and the coded protein will now be functional, thus eliminating the disease. In effect, if the spelling of a gene is wrong and this leads to disease, why not simply change the spelling of the gene to fix the disease? This is the idea behind **gene therapy,** also known as gene transfer, which is the process of inserting new genetic material into an organism in order to overcome an existing genetic defect in that organism.

Methods of Gene Therapy

Multiple gene therapy techniques are being studied, but we will focus here on the most typical method: the use of viruses to transport therapeutic genetic material into target cells within an individual. As shown in figure 13.1, a virus can infect an individual by entering cells and inserting its DNA into the cell nucleus or even into the person's genomic DNA itself. The virus then multiplies by exploiting the DNA replication and transcription machinery of the cell nucleus, thus making more copies of itself; these can then spread to other cells before eventually being targeted by the immune system. This ability to transport DNA into cells makes viruses very attractive from a gene therapy perspective. Through the addition of a "correct" or therapeutic DNA strand to the viral DNA, the virus carries that DNA into the target cells. The cells can then transcribe the therapeutic DNA, resulting in the production of corrected mRNA and protein; thus the correct protein is introduced into the cell and tissue as a disease treatment.

KEY POINT

Gene therapy involves the direct manipulation of the genomic DNA; the intent is to insert a therapeutic DNA sequence, designed to treat a disease phenotype resulting from a genetic mutation.

Gene therapy studies have been conducted (typically in animal models) for a large number of diseases, including Alzheimer's disease, Huntington's disease, and muscular dystrophy; these studies have included experimental trials in humans, but the results have not been as successful as hoped. Because viruses are targeted by the immune system, therapeutic viruses can also become targeted with potentially severe consequences for the patient. Moreover, for viruses that insert their DNA into the genomic DNA of the target cell, there is no predicting where that DNA will insert. If the DNA inserts within an existing gene, instead of within the vast regions of DNA in between coding gene regions, the function of that gene might be disrupted, resulting in other problems for the cell. This mechanism is thought to be the basis for the development of cancers in some children undergoing experimental gene therapy interventions. Finally, the cells of many tissues have a natural life cycle, which means that they die after some period of time. When cells carrying therapeutic DNA die, the therapeutic DNA is destroyed and any advantage to the patient is lost. Thus, many rounds of gene therapy are often necessary in order to maintain an improvement in cell and tissue function.

Controversies Surrounding Gene Therapy

Besides having many technical challenges that have thus far prevented gene therapy from entering the mainstream of medical care, gene therapy is contro-

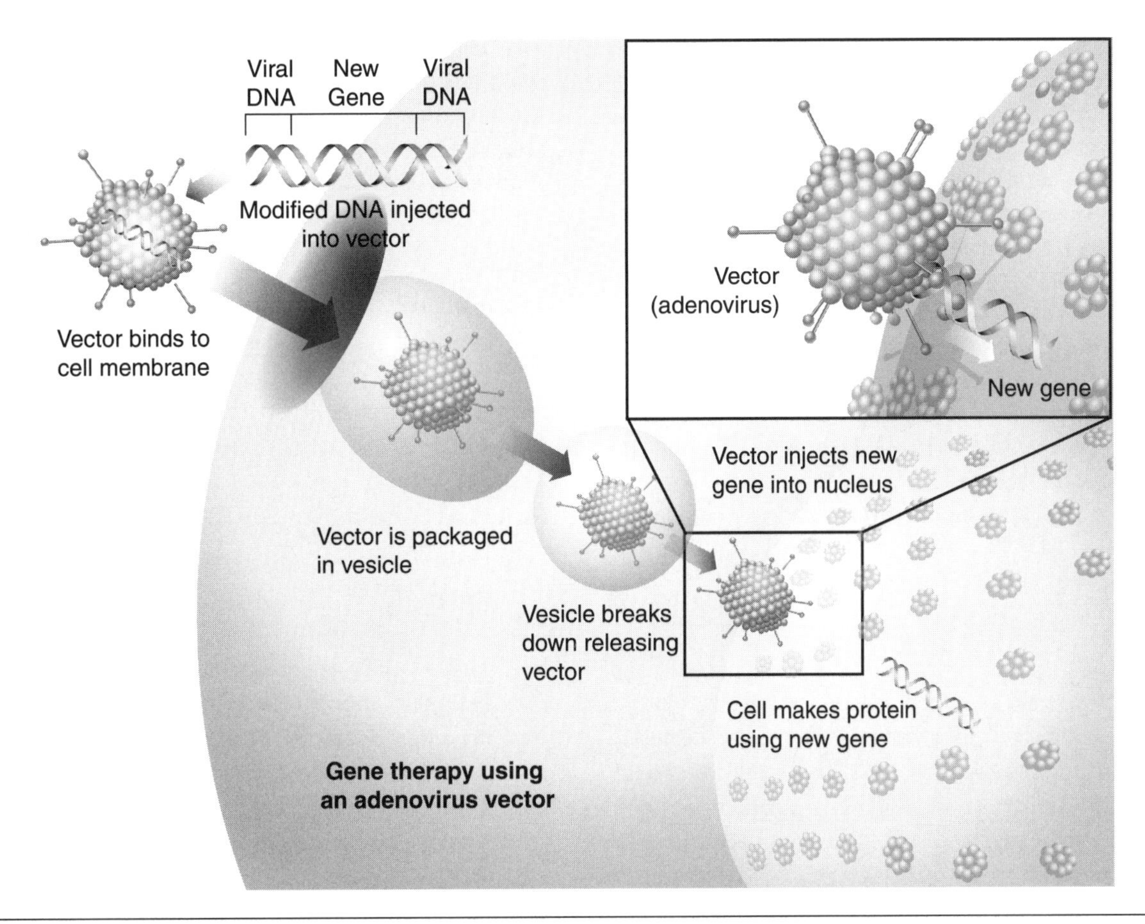

Figure 13.1 An example of gene therapy. A virus (also known as a vector) modified to carry a therapeutic gene sequence is injected into a tissue and absorbed into the target cell. The virus moves into the cell nucleus, where the DNA becomes available for the DNA transcription machinery of the cell, thus allowing the production of a protein from the therapeutic gene sequence. In this way, a mutated or poorly functioning gene within the genome of the patient can be treated by the delivery into the cell of a corrected (therapeutic) DNA sequence.

From the Genetics Home Reference, National Library of Medicine (US). http://ghr.nlm.nih.gov.

versial for two main reasons. First, if gene therapy is performed on germ (i.e., sex) cells, then the genetic manipulation performed in the adult could be passed on to the offspring. Second, what defines appropriate uses of gene therapy in disease treatment? Let's examine these two issues more closely.

In opening this section of the chapter, we discussed gene therapy as a means to erase a genetic defect in a person, most likely one that was tissue or organ specific (e.g., leukemia, muscular dystrophy). Thus, gene therapy would be targeted to the cells in the affected tissue in order to provide new genetic material as a disease treatment. This type of gene therapy is known as **somatic cell gene therapy,** in that these therapies target only developed tissues within an individual. In contrast, gene therapy that might result in the introduction of new genetic material to the sex cells of an individual is known as **germ cell gene therapy.** The germ cells are those that produce the sperm and egg sex cells of the body. Thus, if gene therapy changes the genetic material of the germ cells, then the sperm or egg has the potential to carry that new genetic material to an offspring. In fact, such a change could potentially be passed on to all future generations. Germ cell gene therapy raises concern for many bioethicists in that it has the potential to forever change the human genome sequence, with unknown consequences. Is it appropriate to treat a genetic disorder in an adult's unborn offspring by way of germ cell gene therapy (e.g., treat a dominant genetic mutation)? These questions lead into the second controversial issue surrounding gene therapy: What diseases deserve gene therapy treatment?

While few would argue against the treatment of childhood leukemia, muscular dystrophy, or several other severe genetic diseases with somatic cell gene therapy, few would likely support the use of such technologies to improve intelligence or physical appearance independent of disease. If gene therapy can be used successfully for disease treatment, the same techniques can be applied to **genetic enhancement,** or nondisease therapies. There is no clear-cut boundary between what constitutes a disease deserving of gene therapy treatment versus a physical trait simply treated for genetic enhancement. Does arthritis or obesity qualify as a target for gene therapy, or does hair loss, or eye color? This issue is not unique to gene therapy. Growth hormone therapy has been performed in children likely to have below-average height, even though there was no disease underlying this trait of short stature. Some have argued that this is simply the use of hormonal therapy for physical enhancement and has no place in medicine. The same can be argued for gene therapy, though the number of potential targets for genetic enhancement is considerably larger than for more typical medical therapies. When genetic enhancement is paired with the idea of germ cell gene therapy (i.e., passing along enhancements to multiple future generations), the implications are highly controversial.

KEY POINT 

While gene therapy is generally anticipated as an important technology for the future treatment of rare genetic diseases, its possible use in genetic enhancement or the manipulation of DNA in future generations has led to considerable controversy.

SPECIAL FOCUS
Stem Cells

If you've paid attention to the news media around the world over the past few years, you've certainly heard the phrase "stem cells." In this Special Focus section, stem cells are described from the standpoint of gene therapy, and the reason for their controversial nature is also discussed.

In the process of development, the sperm and egg meet to form a new cell, which ultimately divides many times, forming many cells in the zygote and eventually forming the embryo, and so on. These early cells cannot be distinguished as neurons, muscle cells, skin cells, or any other specific type of cell. In fact, they are remarkable cells in that they have the potential to become any kind of cell needed by the body. These are **stem cells.** In one of the most remarkable aspects of human development, these stem cells divide and the daughter cells eventually become **"differentiated"** such that they are defined more and more as a specific cell type (e.g., neuron, muscle cell). As more and more cells become targeted as a specific cell type, those cells form a tissue built of these similar cells. Once differentiated, it appears that a cell cannot return to its undifferentiated or stem cell state. Thus, as an organism progresses toward adulthood, few stem cells exist in the body, and most of those have become partially differentiated and thus inclined to produce certain types of cells. For example, bone marrow stem cells are predisposed to the production of blood cells, while skeletal muscle stem cells are important for muscle regeneration; these stem cells have become partially differentiated and are inclined to produce only their respective types of cells.

That embryonic stem cells are not differentiated makes them especially appealing from a therapeutic perspective: If a tissue is diseased or damaged in an adult, stem cells could hypothetically be used to generate new, healthy cells for that tissue, thus treating disease. Stem cells have been studied as potential treatments for Parkinson's disease, muscular dystrophy, and many other diseases. But the extent of differentiation in stem cells makes their use controversial. While adults have stem cells, their slight differentiation makes them potentially useful for only a small number of cell and tissue types. **Embryonic stem cells,** on the other hand, are basically undifferentiated, and thus have the potential to be used for the production of any cell or tissue type, drastically increasing their potential in disease therapies.

Unfortunately for scientists, embryonic stem cells must come from embryos, which places stem cell research squarely within the political debate over the beginnings of life and the larger issue of abortion. The process of **in vitro fertilization** (IVF), which is the technique used in infertility treatments of joining sperm and egg outside the woman's body before the fertilized egg is implanted into the womb, can produce many zygotes, some of which are typically discarded after a successful pregnancy. Embryonic

(continued)

SPECIAL FOCUS

(continued)

stem cells have been successfully obtained from such zygotes. Therein lies the controversy: Does the use of an IVF embryo for the development of embryonic stem cells violate the rights of that embryo, or can undeveloped embryos appropriately be used for generating stem cells, which have the potential to save lives of people already born? This is a highly sensitive issue in many countries, so much so that some countries (e.g., Austria, Ireland, Germany, United States) have significantly restricted the use of embryonic stem cells in research. For example, the U.S. Congress has limited the use of federal funds for work with embryonic stem cells to only those cell lines produced prior to mid-2001, though some states (e.g., California) have provided their own funds to support research beyond those federal limits. Many countries, including the United Kingdom, Belgium, Israel, China, and Japan, have few restrictions on the use of embryonic stem cells in health-related research. Some countries, including Australia and Canada, have intermediate policies, with some restriction over the type of research that can be performed using these cells. While research restrictions have arguably slowed research in this area in some countries, considerable progress is being made in understanding the biology of stem cells. Though the true potential of stem cell therapy is not currently known, the anticipated applications of such therapies are remarkable should the ethical dilemmas be overcome.

GENETICS IN SPORT: GENETIC TESTING AND GENE DOPING FOR PERFORMANCE

A natural consequence of the use of genetic information for exercise prescription will be the use of genetic information for sport performance. In fact, the use of genetic information for enhancing sport performance has already begun. In early 2005, an Australian professional rugby team announced that it was using genetic tests of several genes purported to be important for predicting everything from efficiency of oxygen use to muscle strength. Their goal is to use each athlete's genetic information to redesign individual training programs, thus optimizing sport preparation and therefore performance. As discussed previously, the use of genetic tests, especially in early 2005, for enhancing sport performance was likely premature, as the science is still in its infancy. Nevertheless, in the world of athletics, such testing will only expand as individuals and teams seek to improve their performance enough to reach the top of their sport and stay competitive.

Two issues are paramount in the discussion of genetics in sport: (1) the extent to which **genetic testing** (i.e., prescreening of genotype) should be used to predict athletic performance, select and recruit athletes, and optimize training programs, and (2) the use of genetic manipulation via gene therapy to illicitly enhance sport performance. Let's tackle these issues one at a time.

Recall the story from chapter 6 about Eero Mäntyranta, the Finnish cross-country skier who excelled in his sport due to a unique genetic mutation that improved the oxygen-carrying capacity of his blood. Although this a rare case in which the cause of a genetic advantage has been identified, the idea that top athletes are genetically "gifted" compared to lower-performing peers is well accepted (e.g., "He [or she] has good genes"). The difference today is that the basis for genetic advantage in sport performance will be easier to identify, as the genes and alleles underlying sport-related phenotypes are better understood. If Eero Mäntyranta could have been screened for his genetic mutation in the 1960s, certainly he would not have been suspected, as he was, of blood doping to enhance his performance, but would his genetic advantage have been ignored? Would he have been required to compete under different conditions, or banned from competition altogether because of an "unfair" advantage? Should children be screened at an early age, thus allowing parents and coaches to target children into sports in which they are more likely to excel? These philosophical issues underlying sport performance are not easily answerable but are at the heart of the controversy surrounding genetic testing in athletics.

The more challenging questions come from the use of gene therapy to alter genome sequence for sport performance, known as **gene doping.** As discussed earlier, the promise of gene therapy for rare genetic diseases is tremendous, but defining what phenotypes count as disease rather than as a variation of average is not easy. The technology of gene therapy can in theory be applied to virtually any phenotype, including phenotypes related to sport performance. For example, transferring genes for a faster-contracting feline (cat) myosin heavy chain protein (important for muscle contraction), or erythropoietin, or muscle metabolism, would be predicted to result in altered phenotypes with anticipated benefits for various types of sport performance. The consequences of gene therapy for sport performance are unknown and unpredictable, and the use of gene therapy in athletics is more than likely going to fall outside the bounds of evidence-based and ethical science. Any such use of gene therapy would be a modification of a technique designed for experimental disease treatment rather than genetic enhancement or sport performance. Thus, the regulation of gene doping has become a concern for regulatory agencies around the world.

The World Anti-Doping Agency (WADA) in late 2005 announced a series of policies aimed at the use of genetics in sport performance, mostly targeted at the use of gene therapy. While at that time no documented cases of gene therapy for performance had been reported, WADA had already been at work for several years anticipating that some athletes would be unable to resist the temptations of a permanent genetic alteration to improve performance. A major concern for WADA and similar organizations is that gene doping may not be detectable, unlike the illicit use of other drugs and substances to enhance performance. Unlike the situation with a urine or blood sample, how can athletes submit to a muscle biopsy to test for foreign genetic material in the muscle cells prior to competition (muscle cells will be the most likely target of performance-related gene therapy)? Therefore, WADA scientists and policy makers set forth a series of recommendations and policies, including discouraging the use of genetic testing in athletics (with the exception of legitimate medical screening), recommending sanctions for illegal gene doping in sport, and encouraging broader discussions about the dangers and uncertainties of genetic manipulation by both sport organizations and the general public.

Not long after the pronouncements about gene doping were made by WADA, the first indications that gene doping for sport performance may have become a reality surfaced in the news media just weeks before the 2006 Winter Olympic Games. In early 2006, a German coach was on trial for supplying performance-enhancing drugs to athletes. As a part of the information released in the court trial, e-mail correspondence with the coach showed his intent to obtain a gene therapy product known as Repoxygen. Repoxygen is a virus-based gene therapy product designed to deliver the human erythropoietin (EPO) gene to muscle cells for the treatment of severe anemia, a disorder characterized by low red blood cell (RBC) content in the blood. The therapeutic EPO gene was manipulated such that it was expressed during low oxygen levels, thus producing additional EPO, which is the hormone critical for the production of RBCs. Oxford BioMedica, the company that developed Repoxygen, showed it to be successful in the treatment of anemic mice. Repoxygen was never made commercially available for human use because it was far more expensive than standard anemia treatments, which focus on direct injections of the EPO hormone. Moreover, the gene therapy product was never tested in clinical trials in humans before being shelved by the company.

While it was unclear at the time this section was written whether the Repoxygen gene therapy product was ever obtained or used in athletes, the court evidence clearly suggested the intent of the coach to obtain the product for illicit use. Presumably, the gene therapy product would be targeted to skeletal muscle, where it would be available to add to the typical levels of EPO in the blood, thus providing a boost to the normal production of RBCs with consequences for an athlete's oxygen-carrying capacity. An improved oxygen-carrying capacity would have obvious benefits for endurance athletes, though the side effects of excess RBC production can be deadly; this is a concern in this case, since the Repoxygen product has never been tested in humans so its activity in human muscle is not known.

Unfortunately, many anti-doping organizations and scientists think that the Repoxygen example will

KEY POINT

The term "gene doping" refers to the use of gene therapy to enhance sport performance. Though instances of gene doping have not been confirmed, improvements in genetic technology indicate that gene doping is conceptually possible, and the practice has been banned by WADA.

turn out to have been the first of several to reach the headlines over the coming years. Whether or not the EPO produced by the gene therapy product in muscle cells will be different enough from the natural hormone produced in the kidney to allow for testing of athletes is not known. Testing for gene doping is another area of active research for the anti-doping community of scientists. Watch for continued developments in the area of gene doping, as the opportunities for successful genetic enhancement for sport performance will only expand as the technological limitations of gene therapy are overcome.

SUMMARY

The ethical dilemmas surrounding genetics in medicine and sport are extensive. How will DNA be used in medicine while the rights of the patient are respected yet the risks for family members and third parties are acknowledged? Assuming that privacy and legal concerns can be reduced, will individuals willingly seek out genetic testing to identify disease susceptibility? Assuming that the technical difficulties with gene therapy can be overcome, when is its use appropriate? Can the controversies surrounding the use of stem cells in disease treatment be overcome? Finally, how do these issues affect the application of genetic technologies to sport, or should genetics have a place in sport at all? These and other questions hint at the great challenge of genetics in society: The great promise of genetics in medicine and health care is matched by the significant dangers posed by its inappropriate use in nondisease situations.

KEY TERMS

differentiated
embryonic stem cells
ethical, legal, and social implications (ELSI)
gene doping
gene therapy
genetic discrimination
genetic enhancement
genetic testing
germ cell gene therapy
in vitro fertilization
somatic cell gene therapy
stem cell

REVIEW QUESTIONS

1. How is DNA sequence unique from other medical records?
2. In what ways might an individual be hesitant to submit to genetic testing for disease susceptibility or treatment? Similarly, why might potential research subjects be hesitant to participate in a genetics study?
3. How can viruses be used to treat genetic disease?
4. How does somatic cell gene therapy differ from germ cell gene therapy?
5. What is genetic enhancement?
6. What are stem cells and how are they anticipated to be used in medical care?
7. What is gene doping?

APPENDIX

A

Answers to Review Questions

CHAPTER 1

1. An average represents the sum of all measurements, divided by the number of observations, providing a measure of the "typical" or central value across a number of observations. The standard deviation is a measure of the range of individual values that make up the average; see pages 5 and 6.
2. There are three possible sources of phenotype variability, namely: experimental error, environmental factors, and genetic factors.
3. Answers will vary. As an example, in a research study about the phenotype of muscular strength, body size (e.g., height, muscle mass) and previous exercise training history (e.g., strength training) would represent key environmental factors. Other influential factors would be age and sex (see chapter 9 for a complete discussion of these two unique factors).

CHAPTER 2

1. DNA is located within the nucleus of a cell and is organized as 46 total chromosomes (44 autosomes and two sex chromosomes). Males and females differ in their composition of sex chromosomes, with women carrying two X chromosomes and men carrying an X and a Y chromosome.
2. DNA and RNA are made up of long strings of nucleotide bases. In DNA, complementary strands of the nucleotides A, T, G, and C make up a double-stranded helix structure. In contrast, RNA is single stranded, containing the nucleotides A, G, C, and U (uracil, replacing T, thymine, found in DNA).
3. DNA is used as the instructional information to form a complementary mRNA strand, which is then used to manufacture the coded protein. These processes are known as transcription and translation, respectively.
4. The exons or exonic sequences of the gene, found in the coding region, contain the information for the amino acid sequence. The promoter, introns, and terminator regions contain regulatory sequences important for gene function.

CHAPTER 3

1. Transcription is the process of reading a DNA sequence and producing the corresponding mRNA sequence by RNA polymerase. This process takes place in the nucleus of the cell, and the entire coding region of the gene (between the promoter and terminator regions) is transcribed.
2. Translation is the process of reading the mRNA sequence and manufacturing the corresponding amino acid sequence by the ribosome. This process takes place in the cell cytoplasm after the mRNA has undergone posttranslational modification, whereby the intronic sequences are removed and the exon regions are spliced together.
3. The ribosome recognizes the AUG initiation codon within mRNA as the start site for the translation of amino acid sequence. Translation ends when the ribosome reaches a termination codon, of which three are possible (UGA, UAA, UAG).

4. The genetic code is the universal rule by which the ribosome translates mRNA sequence into the corresponding amino acid sequence. Each three-letter codon, beginning with the initiation codon (AUG), codes for a specific amino acid, a rule that is unchanging across all cell types. Of the 64 three-letter combinations, three code for termination codons while all others code for one of the 20 amino acids.

CHAPTER 4

1. All 46 chromosomes are contained within a somatic cell, and those same 46 (exact duplicates) are contained in the two daughter cells after mitosis and cell division of the original somatic cell.
2. All 46 chromosomes are contained within the original germ cell prior to meiosis and cell division. The end result of gamete formation is the formation of four gametes, each with 23 chromosomes (1-22 autosomes and one sex chromosome).
3. Chromosomal recombination occurs during meiosis and consists of two processes that ensure the shuffling of genetic information for the offspring: crossover and independent assortment. Crossover is the process whereby homologous chromosomes align and exchange DNA sequence, while independent assortment is the random separation of different chromosomes into the resulting gametes.
4. Chromosome recombination and inheritance patterns in women are the same for the sex chromosomes as they are for the autosomes, as women carry two X chromosomes. In contrast, because males carry one X and one Y chromosome, crossover cannot occur during meiosis. Thus, a male's female offspring will inherit the father's unchanged X chromosome and male offspring will inherit the father's unchanged Y chromosome.
5. All offspring would inherit at least one copy of the recessive disease gene, but only 50% of offspring would inherit two copies, thus being affected by the disease.

CHAPTER 5

1. We can determine heritability by measuring phenotype values in both monozygotic and dizygotic twin pairs. If genetic factors are important for a trait (i.e., the trait is heritable), the correlation between phenotype values in monozygotic twins will be greater than that seen in dizygotic twins, and heritability can be estimated by the equation $h^2 = 2(r_{MZ} - r_{DZ})$.
2. Genetic variation is any position within the genome sequence where more than one nucleotide is observed in a population. The most common type of genetic variation is single-nucleotide polymorphisms, in which a single nucleotide differs among individuals. Other types of variation include insertion/deletion and repeat polymorphisms.
3. A mutation and a polymorphism both represent genetic variation at a specific position within the DNA; however, a mutation is a genetic variant for which the rare allele is found in only a very small number of individuals (as few as one), while for a polymorphism the rare allele is found in at least 1% of the population.
4. DNA sequence variation can influence the function of a gene by altering regulation of that gene, for example by affecting transcription factor binding sites in the promoter or other regulatory region of the gene sequence. If genetic variation occurs within the coding region of the gene, that variation could alter the amino acid sequence of the resulting protein, with the potential for altered protein structure or function. An altered protein function could then affect the phenotype associated with that protein.
5. Answers will vary. For example, a possible nonsense mutation would be one in which the coding strand of DNA contained a SNP such that the two sequences were TAT and TAA, where TAT codes for tyrosine and TAA is a stop codon (A/T SNP in the third position of the codon). A possible missense mutation would be the two sequences CTC and CCC, which code for leucine and proline, respectively (T/C SNP in second position of the codon). A possible silent SNP would be the two sequences GGT and GGA, both of which code for glycine (T/A SNP at third position of the codon).
6. Mutation in a DNA sequence can occur through a variety of different mutagens (e.g., X-rays, chemicals). If this mutation occurs within the germ-line cells of an individual, and the mutation is then passed along to the offspring of that individual, the DNA sequence has undergone

evolution. Evolution is simply a change in DNA sequence in succeeding generations in a population. If this mutation has either a neutral or a positive effect on evolutionary fitness, then the mutation may spread in a population to become a polymorphism.

CHAPTER 6

1. Mendelian traits are typically governed by a single gene or gene mutation, without the influence of other factors (e.g., environmental factors). Complex traits, in contrast, are influenced by many genetic factors and many environmental factors.
2. This disease has a 90% penetrance. Some disease gene carriers could be spared disease because of other unique genetic variants carried in their DNA (i.e., gene × gene interactions) or unique environmental factors (i.e., gene × environment interactions).
3. An intermediate phenotype is one that underlies a broader phenotype; for example, hemoglobin levels would represent an intermediate phenotype for the broader, complex phenotype of $\dot{V}O_2max$. Intermediate phenotypes are easier to study in that they are influenced by fewer genes and environmental factors; however, such phenotypes are more challenging in that they often require more sophisticated or invasive techniques for phenotype measurement.
4. Answers will vary. As an example, body height could be influenced by maternal nutrition during pregnancy, nutrition during childhood and adolescence, medication use, and various diseases (socioeconomic status correlates with body height!). Genetic factors are assumed to contribute as much as 80% of the variance in body height, but many genes may contribute. Potential intermediate phenotypes could be anatomical characteristics of the bone growth plates at certain joints, bone density, and blood markers of bone metabolism.
5. A gene × environment interaction is one in which an environmental factor (e.g., physical activity) can alter the typical influence of a genetic variant (e.g., ε4 allele of the ApoE gene), such that the typical genetic disease predisposition (e.g., Alzheimer's disease) is altered in the presence of the environmental factor.

CHAPTER 7

1. Linkage disequilibrium describes two polymorphisms or alleles that tend to travel together and thus be predictive for one another during recombination. Because one allele in the first polymorphism is informative for the linked allele in the second polymorphism, it can serve as a "marker" allele for that second polymorphism.
2. A haplotype block is a stretch of DNA sequence that tends to travel as a single unit of material during recombination events; it is bounded by recombination hotspots. Most haplotype blocks have several spellings, with three to five haplotype sequences being common in a population.
3. A linkage study uses hundreds of marker polymorphisms to determine correlations between specific genomic regions and a particular trait; such studies are performed in families, with each member genotyped for these linkage markers and also measured for the trait of interest. The resulting linkage map will show regions of high LOD scores, which are genomic regions correlated with the phenotype. These regions can then be studied more intensively to identify functional genes and polymorphisms important to the phenotype.
4. Haplotype blocks contain many SNPs, and some of these SNPs will be in complete linkage disequilibrium (completely redundant). Thus, when one is determining haplotype for a particular block, only nonredundant SNPs, known as "tag SNPs," are required in order to identify possible haplotypes.
5. Diplotype refers to the combination of haplotypes carried for the maternal and paternal chromosomes of an individual, much as genotype describes the two alleles carried by an individual. (a) Diplotype is G-A-T and G-A-G. (b) Diplotype is G-A-T and C-A-T. (c) Diplotype cannot be determined (multiple heterozygous genotypes).
6. Epigenetic factors are heritable factors that are independent of DNA sequence variation, including DNA methylation labels on DNA nucleotides. DNA methylation, when present within a regulatory region of a gene (e.g., promoter), can alter transcription of that gene, which could potentially alter the amount of protein produced by the gene.

CHAPTER 8

1. The critical question is, Does the phenotype of interest have a genetic component (is it heritable)? If not, there is no rationale for pursuing a genetics study for that trait.
2. The search terms include gene, genotype, allele, polymorphism; the complete list is shown on page 79.
3. The key priority for identifying novel candidate genes is to focus on the physiology underlying the phenotype of interest. In general, those proteins that are important to the function of the phenotype, and are also very specific to that phenotype, will rise to the top of your priority list of potential candidate genes (e.g., a gene specifically expressed in muscle when your phenotype is muscle mass).
4. Though answers will vary, the text has focused on the use of the NCBI Web site for identifying the official gene name and symbol, as discussed on page 82.

CHAPTER 9

1. See the Key Point on page 88. Association studies focus on the recruitment (if necessary) of specific genotype groupings, while case–control studies focus on phenotype categories. One can develop a case–control study from an association study by selecting out the individuals with the highest and lowest values of a phenotype, thereby "building" case and control groups.
2. Low rare allele frequencies can require prescreening for genotype, grouping by allele, or skipping a particular polymorphism for study, all of which can cause problems for a research study. The expected genotype frequencies are 64%, 32%, and 4%.
3. In brief, genetic variation does differ slightly by race, as do various health disparities. A full discussion can be found in the Special Focus section within the chapter.
4. A primary concern is not having a balanced sample size for polymorphisms with a low rare allele frequency. Moreover, a retrospective study often will not have taken into account all of the necessary environmental variables that would be optimal to maximize the ability to identify a genetic association.
5. Cause-and-effect conclusions are not typical from human genetics studies because we rarely manipulate the genetic complement. Rather, we most often study genetic correlations—how subjects in different genetic groups differ. Animal models, which allow genetic manipulation, can provide a means of addressing cause and effect.

CHAPTER 10

1. A cheek cell or buccal cell sample, collected via swab or mouthwash solution, can be used to collect a DNA sample.
2. The first step is denaturation, which is the separation or melting of double-stranded DNA. The second step is annealing, during which the primers bind to the single-stranded DNA. The third step is extension, when the polymerase enzyme copies the single-stranded DNA material downstream of the primer. The end result, after several cycles of PCR, is millions of copies of the fragment of DNA defined by the PCR primers.
3. A restriction enzyme cuts double-stranded DNA at a very specific sequence, usually of four to eight bases in length. Through selection of an enzyme that cuts one allele of a polymorphism but not the other allele, PCR fragments containing the polymorphism can be subjected to enzyme digestion, and the resulting fragment lengths can be used to identify the genotype for each sample.
4. Fluorescent tags can be added to allele-specific probe sequences that will bind directly to the polymorphism of interest. With a unique fluorescent tag used for each allele-specific probe, the fluorescence emitted from each sample can be used to determine genotype.
5. DNA sequencing is primarily used to identify the exact ordering of nucleotide bases in a particular region of DNA. For genetic association purposes, sequencing can also be used to identify polymorphisms within the sequenced region by observed sequence differences among multiple samples. Individuals heterozygous for a particular polymorphism will show a unique chromatogram waveform at the location of the polymorphism.

CHAPTER 11

1. Because ACTN3 is present only in type II skeletal muscle fibers, which are important for generation of muscle power used in sprint

or power activities, the absence of the protein in X/X homozygotes was predicted to have an adverse influence on performance in those events.

2. Two nearby missense polymorphisms in the ApoE gene result in three possible alleles, ϵ2, ϵ3, and ϵ4. The ϵ4 allele has been consistently associated with increased risk of late-onset Alzheimer's disease, with the greatest risk occurring in ϵ4 homozygotes.
3. Although the data are inconclusive, several studies have shown a positive association between the I allele and endurance performance, but such an association with $\dot{V}O_2max$ has not been consistently observed. Researchers are now beginning to think that the link between the I allele and endurance performance could be through muscle or metabolic efficiency, perhaps through an influence on muscle fiber type, though much work remains to be done in order to verify that hypothesis.
4. Yes. A rare myostatin gene mutation has been reported in a single child that appears to result in enhanced muscle mass growth. The bulk of human studies, however, suggest that common polymorphisms in the myostatin gene have only a small influence on skeletal muscle mass phenotypes in the general population.
5. Answers will vary.

CHAPTER 12

1. Personalized medicine refers to the incorporation of genetic information into the health care of an individual. Specifically, health care interventions will be individualized to a person's specific genetic profile, thus increasing efficacy and reducing side effects.
2. Pharmacogenomics is the specific application of genetic information to both the development of novel prescription drugs and the prescription of existing drugs. The TPMT gene test example shown on page 131 is an example of a gene × drug interaction that is relevant for gene screening prior to drug prescription.
3. Specific examples will vary. In general, a specific genetic profile may be more or less responsive to a diet or exercise intervention designed to reduce disease risk. Knowing this information (through gene screening) may improve disease treatment (or reduce risk) by providing the ability to individually tailor the diet or exercise prescription to the genetic profile. An example is the ApoE ϵ4 allele, which increases risk for Alzheimer's disease; this risk appears to be reduced in physically active people. Thus, exercise prescription would be highly recommended in carriers of the ϵ4 allele (as discussed in chapter 11).
4. Personalized disease prevention will necessitate gene screening in advance of disease onset, which will require individuals to seek out such screening. Moreover, the genes and physiological pathways important to prevention of a disease may differ from the genes and pathways important for disease treatment.

CHAPTER 13

1. Unlike most clinical measures included in a medical record, DNA sequence is unchanged over time, can be predictive of disease, and is shared among family members.
2. The primary concern of individuals undergoing genetic testing is possible genetic discrimination in employment or insurance access or both. Will such information be kept private and secure so that any disease susceptibility will be not be used against people in the future? Research participants have similar concerns, though in some cases participants may view not receiving genetic information as a negative.
3. Viruses can easily insert themselves into a cell nucleus (or even into genomic DNA) and then exploit the DNA replication machinery of the cell to make copies of themselves. Inserting a therapeutic gene sequence into a virus and then inserting the virus into the target (i.e., diseased) tissue can potentially restore the tissue's function and treat the genetic disease
4. Somatic cell gene therapy works through the insertion of therapeutic DNA material into the somatic (nonreproductive) cells of the body. Such changes cannot be passed on to future generations. In contrast, germ cell gene therapy affects the genomic DNA of the germ cells; the therapeutic DNA can then be transmitted into sperm or egg, thereby permitting transmission of the therapeutic DNA sequence to future generations.

5. Genetic enhancement is the use of gene therapy to alter a nondisease phenotype (i.e., intelligence, physical appearance).
6. Stem cells are a cell population that is not completely differentiated but rather has the potential to differentiate into a number of different cell types. Stem cells can hypothetically be used to generate new, healthy cells for a diseased tissue, thus treating the disease.
7. Gene doping is the anticipated use of gene therapy for the enhancement of sport performance.

APPENDIX

B

Evolution and Hardy-Weinberg Equilibrium

If evolution is simply a change in DNA sequence over time (as discussed in chapter 5), then what are the forces that can result in such changes? Four major forces that can act on a species are capable of influencing DNA sequence over time: **mutation, natural selection, migration,** and **genetic drift.** The first two of these are discussed in chapter 5: (1) *mutation* of DNA, which can occur in a single individual through simple DNA copying error or through some environmental mutagen, with the resulting new allele possibly passed along to subsequent generations if it is present in the sex cells; and (2) *natural selection,* the concept that any new allele that provides some survival or reproductive advantage for an individual will tend to spread among the population over many generations. The other two evolutionary forces are also straightforward. (3) *Migration* is the idea that some small subgroup of individuals from a population leaves that population, for whatever reasons, and forms a new population in another environment. The new population may be alone in the new environment, or may interact and mate with another unique population in the area. The new subgroup population's DNA sequence, while obviously similar to that of the population from which the subgroup came, will be uniquely derived from the alleles carried by the migrating subgroup's DNA sequence, or the subgroup's DNA sequence in combination with that of another unique population in the new environment. Thus, when small subgroups migrate and form new populations, DNA sequence changes can occur, resulting in slight phenotype differences in the new populations over time. In humans, the various races and ethnicities across the world are thought to have developed from migratory subgroups that left an original population, which was based in Africa. Finally, (4) *genetic drift* is a concept similar to migration, but more random in nature. Over time (i.e., generations), various alleles may increase or decrease in frequency (i.e., "drift") in the genome simply by chance, especially in small populations. These allele frequencies are changing not due to natural selection, but rather due to the randomness of mate selection and reproductive chance that occurs among individuals. For example, the frequency of the A allele at an A/T polymorphism increases from 25% to 40% in a small population over a few generations, though there is no selective advantage associated with the A allele. This random change in allele frequency most often occurs in very small populations, in which individuals have limited reproductive choice. So, all four of these forces can result in changes in DNA sequence in a species over time, which is evolution.

Why spend so much time discussing this background on evolution? The reason is that one of the questions geneticists ask is whether or not evolutionary forces (i.e., these four) are acting on a particular DNA sequence. And one way of addressing this question is to examine allele and genotype frequencies, which are the end result of evolutionary change. Two individuals working independently, G.H. Hardy and W. Weinberg, developed a way of testing for the involvement of evolutionary forces by asking what would happen to DNA sequence over time if *none* of these evolutionary forces were in action. In other words, what would happen to allele and genotype frequencies over time if there were no mutation, natural selection, migration, or genetic drift in a population, and all individuals of the species mated randomly? The answer is that DNA sequence would not change, as there are no forces acting to cause a change! Now, this is not a real-world scenario, but rather a hypothetical condition that is useful for testing the presence or absence of evolutionary forces on a particular DNA sequence. This test, now known as a test of **Hardy-Weinberg equilibrium** or the *Hardy-Weinberg law*, is often performed in genetic association studies. Let's first see how it works, and then we'll discuss the use of the test in the real world.

We'll perform the Hardy-Weinberg equilibrium (HWe) test on a hypothetical single-nucleotide polymorphism, with alleles A and B, at some location in the genome. The HWe law states that random sexual reproduction alone cannot change the frequency of alleles (A/B) or genotypes (A/A, A/B, and B/B) in a population over time, assuming that none of the four forces of evolution are acting on the population. The HWe test involves some simple math. Here are the variables we'll use:

Frequency of the A allele = p.

Frequency of the B allele = q.

Since A and B are the only possible alleles, then the frequencies of A and B add to 1:

$$p + q = 1.$$

If we know the proportion of one allele in a population, we can calculate the other allele:

$$p = 1 - q \qquad \text{or} \qquad q = 1 - p.$$

Where do we get the allele frequencies? The frequencies come from the genotype data of our research study subjects. Now, we predict the genotype frequencies that would result from these allele frequencies under the assumption of the HWe law, as follows:

$$p^2 + 2pq + q^2 = 1$$

where p^2 is the frequency of A/A individuals, $2pq$ is the frequency of the heterozygous (A/B) individuals, and q^2 is the frequency of B/B individuals.

Once the expected genotype frequencies have been calculated, they can be compared to the actual genotype frequencies in our sample. Are our actual genotype frequencies different from the frequencies expected under Hardy-Weinberg equilibrium conditions? In other words, do evolutionary forces appear to be at work on the tested polymorphism? This comparison is statistically tested using a chi-square test. An example of the entire calculation is shown in figure 1.

Any deviations from the expected Hardy-Weinberg equilibrium values *can* represent evolutionary forces at work in the population, because the genotype frequencies of our actual population are not what we would expect to have in the absence of evolutionary forces. I emphasize the word "can" because there are other reasons why the calculation could give you a deviant result (i.e., out of Hardy-Weinberg equilibrium), none of which are important from an evolutionary perspective. One of the biggest sources of error in the HWe calculation is a small sample size that is not representative of the population as a whole. If the actual genotype values going into the calculation are not truly representative of the entire population from which that sample was obtained, then the test is invalid. Moreover, if the population contains individuals from multiple geographic ancestries (e.g., different races), a topic discussed fully in chapter 9, the genotype data may be skewed from what would be seen in any one of those subpopulations, thus giving an invalid result. All of this is to say that using the HWe test as a measure of evolutionary forces at work is problematic unless the sample is very large and homogeneous. In fact, many researchers have taken to using the HWe test as a way to demonstrate adequate sample size and accurate genotype data, neither of which the test was designed to show, and some have called for an end to the HWe test unless researchers are specifically looking to address the question of evolutionary forces at work.

So, what is the bottom line for folks gaining an introduction to the field of genetics? Unfortunately, this is a controversial area with arguments on both sides. Ultimately, you will see the HWe test performed in many genetic association studies, and it may be requested for any work that you might submit for publication even if your questions of interest do not revolve around evolutionary forces (hence, the need for this big discussion in an introductory text!). If your results do deviate from HWe expectations, the most likely reason is either small sample size or a heterogeneous sample. In this scenario, my recommendation would be to verify genotyping accuracy (a potential source of error in the calculation) and then determine if your genotype frequencies are similar to those shown by other researchers or in polymorphism databases. If your genotype frequencies differ from those

Figure 1 Calculating the Hardy-Weinberg Equilibrium

Calculating Hardy-Weinberg Equilibrium (HWe) is fairly straightforward. Let's calculate it with an easy example:

1. First, determine the number of individuals with A/A, A/B, and B/B genotypes in your sample.

   50 A/A, 25 A/B, 25 B/B

2. Now, calculate the number of A and B alleles in the sample. Recall that each genotype represents 2 alleles, so each A/A genotype represents 2 A alleles, and so on.

   $$\text{Number of A alleles} = (\#A/A \times 2) + (\#A/B \times 1) = (50 \times 2) + (25 \times 1) = 125$$

   $$\text{Number of B alleles} = (\#B/B \times 2) + (\#A/B \times 1) = (25 \times 2) + (25 \times 1) = 75$$

3. What are the frequencies of the A and B alleles? Divide the values calculated in step 2 by the total number of alleles in the sample (2 alleles per person). This calculation gives you the values for p and q.

   $$\text{Total alleles} = \text{total number of individuals} \times 2 = (100 \times 2) = 200$$

   $$\text{Frequency of A alleles} = p = 125/200 = 0.625$$

   $$\text{Frequency of B alleles} = q = 75/200 = 0.375$$

4. Now, we can calculate the genotype frequencies expected under the conditions of Hardy-Weinberg equilibrium by calculating p^2, $2pq$, and q^2.

   $$p^2 = (0.625 \times 0.625) = 0.39\ (39\%)$$

   $$2pq = (2 \times 0.625 \times 0.375) = 0.47\ (47\%)$$

   $$q^2 = (0.375 \times 0.375) = 0.14\ (14\%)$$

5. How many people are predicted for each genotype, assuming the same total sample size? Multiply the values in step 4 by the number of samples in the research study.

   $$\text{Expected number of A/A individuals} = (0.39 \times 100) = 39$$

   $$\text{Expected number of A/B individuals} = (0.47 \times 100) = 47$$

   $$\text{Expected number of B/B individuals} = (0.14 \times 100) = 14$$

6. Now the question is whether the number of individuals predicted for each genotype by the HWe test is different from the number of individuals actually carrying each genotype in our research study. To test this, we'll use the chi-square test. I'll skip the details of the chi-square test, which can be found in any introductory statistics textbook, with one exception. When performing the test, the number of degrees of freedom (df) is equal to 1, because the test is really based on the number of alleles (2 alleles – 1 = 1 df). For this example, the chi-square value is 20.2, which represents a statistically significant difference between the expected and actual values ($P < 0.001$ with 1 df). Therefore, we would reject the null hypothesis that the population is in Hardy-Weinberg equilibrium. Now, we have to figure out why our frequencies deviate from expected HWe, as discussed in the main text.

reported by others, could this be due to heterogeneity in your population or to a small, nonrandom sample? Addressing these issues is most important in discussion of the results of a HWe test. I would recommend against discussing possible evolutionary forces at work in your sample unless this is involved in one of the hypotheses you proposed at the beginning of the study.

KEY TERMS

genetic drift
Hardy-Weinberg equilibrium
migration
mutation
natural selection

Glossary

additive—Referring to a gene or allele that is considered neither dominant nor recessive in its contribution to a phenotype. Multiple additive alleles each contribute proportionately to the phenotype.

adenine (A)—A nucleotide base found in both DNA and RNA. Adenine always binds with thymine (T).

admixture—The extent to which an individual's DNA sequence is based on different geographic ancestral origins.

agarose—A seaweed-derived compound used in the manufacture of gels employed in gel electrophoresis.

allele frequency—The extent to which a specific allele is present in a population.

alleles—Genes or nucleotides that are located at the same position on two corresponding or homologous chromosomes. Two alleles exist for every gene or nucleotide for all autosomes, and for the X chromosome in females.

alternative splicing—A process whereby exon regions within an mRNA sequence are rearranged or removed in order to form a unique protein sequence.

amino acid—An organic molecule that acts as a building block for the manufacture of proteins.

annealing—The process of two complementary DNA or RNA sequences binding together. In the second phase of PCR, primers anneal to the DNA template sequence.

association study design—Referred to in this text as a research method used to identify phenotype differences among different genotype groups for a particular polymorphism; subjects are recruited by genotype rather than by phenotype.

autosome—Any chromosome that is not a sex chromosome (X or Y). There are 22 autosomes in the human genome.

average—the value regarded as typical of a group of numbers, calculated by adding all of the numbers together and dividing by the total amount of numbers.

bioinformatics—The use of computational and statistical techniques to address complex biological research questions; also known as computational biology.

buccal cell—A cell of the inner lining of the cheek that can be used for DNA collection.

buffy coat—The layer of white blood cells that separate after centrifugation of a whole-blood sample. These white blood cells are used for DNA collection.

candidate gene—A gene hypothesized to have an influence on a trait of interest, assuming that genetic sequence variation exists within that gene.

case–control design—A comparison study design typically examining subjects or patients with a condition and otherwise matched subjects without that condition (e.g., disease).

cell division—The process by which a single cell splits into two daughter cells. The two types of cell division are mitosis and meiosis.

chiasma—The breakpoint at which two paired homologous chromosome copies exchange genetic material during crossover.

chromatogram—For this book, the output of a DNA sequencing reaction using fluorescent tags showing each nucleotide as a wave in a series corresponding to the exact sequence of nucleotides in the DNA sequence.

chromosomal recombination—The process by which the genetic material in two parental chromosomes becomes shuffled during meiosis to result in a novel DNA sequence in the resulting germ cells. Recombination can occur via the processes of crossover and independent assortment.

chromosome—A distinct, long strand of DNA sequence with associated packaging proteins within the genome. There are 46 total chromosomes in the typical human cell.

coding region—A region within the DNA sequence of a gene that contains information important for the manufacture of a protein.

codon—A DNA sequence of three nucleotides (a triplet) within an exon of a gene that corresponds to a specific amino acid or a stop signal for the translation process.

complementary base pairs—The DNA nucleotides bind to each other in a specific pattern, such that adenine (A) binds only to thymine (T), and guanine (G) binds only to cytosine (C) in the DNA double-stranded structure. The RNA nucleotide uracil (U) binds only to adenine (A) during the process of transcription.

complex disease—A disease phenotype that is influenced by both genetic and environmental factors, and likely the interaction of genetic and environmental factors. Thus both genetic and environmental susceptibilities can be identified as contributing to disease risk.

correlation—The extent to which two or more variables are similar.

crossover—The process of exchanging DNA sequence regions among two paired, homologous chromosomes during meiosis.

cytosine (C)—A nucleotide base found in both DNA and RNA. Cytosine always binds with guanine (G).

D allele—Refers to the deletion allele of an insertion/deletion polymorphism.

denaturation—The process whereby two complementary bound strands of DNA or RNA separate in response to high temperature. The first step of PCR is denaturation of the double-stranded DNA to single-stranded DNA.

deoxyribonucleic acid (DNA)—Combinations of nucleic acids in series, the exact sequence of which contains information required for the production of proteins and other cell components. DNA consists of the nucleotides adenine, guanine, cytosine, and thymine.

differentiated—A cell that has developed unique properties specific to its function within a tissue, as opposed to a stem cell, which is considered free of such unique properties.

diplotype—The combination of two haplotype sequences carried by an individual. Diplotype is similar to genotype in that it describes the combination of the two haplotype alleles.

dizygotic twins—Two individuals conceived at the same time but developed from different fertilized eggs. The individuals are born together and share the uterine environment, but they share only 50% of their DNA sequence variation as is the case for other siblings. Dizygotic twins differ from monozygotic twins.

DNA extraction—The process of separating DNA material from other cellular components; often performed on white blood cells or buccal cells.

DNA methylation—The presence of methyl groups added to particular DNA nucleotides, especially cytosine, which can be inherited without changing the DNA sequence.

DNA polymerase—The enzyme responsible for the copying of a DNA sequence during cell replication or the PCR assay.

DNA primers—Short, single-stranded fragments of DNA sequence used in the polymerase chain reaction (PCR).

DNA replication—The copying of the genetic information in a cell in preparation for cell division, or the production of two individual cells.

DNA sequencing—The process of determining the exact order of nucleotides in a specific region of DNA.

dominant—Referring to that allele of a polymorphism that determines the phenotype to the exclusion of the other allele, despite the presence of both alleles in the cell. Only one copy of a dominant disease allele is required for disease to develop. Dominant is the opposite of recessive.

elongation—See extension.

embryonic stem cell—An undifferentiated cell taken from an embryo that has the potential to develop or differentiate into multiple cell types.

environmental factors—Stimuli considered outside the body or independent of DNA sequence that affect a phenotype. Smoking is an environmental risk factor for lung cancer.

epigenetics—The study of meiotically heritable differences in gene function that are independent of changes in DNA sequence.

error variability—See experimental error.

Ethical, Legal, and Social Implications (ELSI)—A subgroup of the National Human Genome Research Institute dedicated to studying the ethical, legal, and social implications of the complete sequencing of the human genome.

evolution—Any change in DNA sequence that is maintained in a species over multiple generations, with or without any influence on a trait.

exon—A region of DNA sequence within a gene that carries information important for the production of a protein. Exons are composed of codons and separated by introns. Complete protein sequence information is typically contained in multiple exons in a gene.

experimental error—Variation in the measurement of a trait that is due to technical problems resulting in inaccurate or imprecise measurements in an individual. Experimental error can contribute to trait variability; thus it must be minimized in any study of the influences of environmental and genetic factors on a trait of interest.

extension—The process whereby a polymerase enzyme adds nucleotides in sequence during the process of DNA or RNA copying. The third step of the PCR assay involves the elongation of the DNA primer to make a new copy of a DNA strand. Extension is also known as elongation.

familial aggregation—Similarity of trait or phenotype measurement values in related individuals. Also known as familial resemblance or similarity.

fitness—In the context of evolution and natural selection, the ability of an individual to survive to adulthood and reproduce in comparison to that individual's peers. Fitness is also used separately from the context of evolution to describe an individual's ability to perform physical activity.

fluorescent tag—A molecule that contains a particular light-emitting property that allows it to be used to identify particular nucleotides in DNA sequencing and genotyping reactions.

frameshift mutation—An allele, especially an insertion or deletion of one or more bases in a coding region of a gene, that shifts the DNA sequence. An allele not divisible by 3 will result in the insertion/deletion of a partial codon, thus shifting the sequence of triplets and altering the protein sequence downstream of the frameshift allele.

fraternal twins—See dizygotic twins.

gametes—The sex cells, either egg (ovum) or sperm, which combine to form a zygote in the process of human reproduction. Each sex cell contains half the complement of DNA sequence required for development.

gel electrophoresis—A laboratory method that allows the separation of different lengths of DNA sequence by the mobility of DNA strands in an electric field.

gene—A region of DNA that encodes or contains the information required to make a protein. The gene region is often described as the coding region of the DNA sequence plus the nearby regulatory regions needed for transcribing the DNA sequence.

gene doping—The use of gene therapy to illicitly improve sport performance through the incorporation of novel genetic material into the cells of an athlete; similar to drug doping for performance enhancement.

gene expression—An alternative term used to describe gene transcription, the process of reading DNA sequence and manufacturing a complementary mRNA sequence.

gene region—See gene.

gene therapy—The insertion of therapeutic DNA sequence into a cell or tissue for the purpose of correcting a disease resulting from a mutation in genetic sequence.

genetic code—The set of rules that specify which three-nucleotide codons (triplets) correspond to which amino acids or termination signals in the mRNA sequence.

genetic discrimination—The use of DNA sequence information, especially DNA mutations and genotypes, to unfairly treat a particular individual or group of individuals.

genetic drift—An evolutionary force resulting in a change in DNA sequence due to random chance, often associated with reproduction in small populations.

genetic enhancement—The use of gene therapy to alter cell or tissue function for non-disease purposes.

genetic factors—Heritable DNA sequence information that affects a phenotype. Genetic variation among individuals that affects disease predisposition, independent of environmental factors, is an example of a genetic factor.

genetic profile—Used to describe the entire complement of DNA sequence variation carried by an individual, which will be unique from all other individuals (except identical twins).

genetic testing—Screening DNA sequence or polymorphism information to predict disease susceptibility or phenotype characteristics.

genetic variation—Differences in DNA sequence that exist at the same nucleotide position among different individuals; in other words, when the spelling of a gene region differs between two individuals.

genome—The entire complement of DNA sequence and genes contained within a typical cell. The human genome consists of 3.1 billion nucleotides found over 22 autosomes and 2 sex chromosomes with 20,000 to 25,000 genes.

genome database—A collection of DNA sequence and related genetic information for a particular organism. Several such databases are available to provide full and free access to the human genome sequence as well as the genomes of many other species.

genotype—The combination of two alleles present within a single individual, corresponding to the two alleles present on the paired chromosomes. Genotype also refers more generally to the genetic profile or genetic makeup of an individual.

geographic ancestry—Refers to the region in which a particular subpopulation has lived for many generations (e.g., Northern Europe or East Asia).

germ cell—A precursor cell that develops into a gamete, or sex cell, and is thus responsible for the formation of new offspring. Germ cells are distinct from somatic cells.

germ cell gene therapy—The use of gene therapy on germ cells; a possible consequence is the transmittal of the therapeutic gene product to future generations.

guanine (G)—A nucleotide base found in both DNA and RNA. Guanine always binds with cytosine (C).

haplotype—A specific combination of neighboring alleles from different polymorphisms that tend to be inherited together. A haplotype can be thought of as an allele for a larger region of DNA, known as a haplotype block, such that different haplotype spellings are seen for the same haplotype block, just as different alleles are seen at a specific polymorphism.

haplotype block—A region of DNA sequence that is inherited together during recombination, whose boundaries are defined by recombination hotspots where crossovers tend to occur during meiosis. Haplotype blocks are distinguished from recombination blocks in that they are population-level recombination events rather than family-level recombination events.

HapMap—A genome-wide map describing the location of typical haplotype block boundaries and haplotype sequences in the human genome.

Hardy-Weinberg equilibrium—The theory that DNA sequence will not change over many generations if no evolutionary forces are present and mating is random among individuals in a population.

heritability—That portion of variation in a phenotype due to genetic factors, and thus inherited by offspring from parents. Heritability is a quantitative estimate of the contribution of genetic factors to a trait of interest.

heterozygote—The term used to describe an individual carrying two different alleles at a particular polymorphism in both gene copies.

homologous chromosomes—An identical chromosome pair, as is seen during the crossover process of chromosome recombination. Four copies of the same chromosome align perfectly in preparation for crossover.

homozygote—The term used to describe an individual carrying two identical alleles at a particular polymorphism in both gene copies.

human genome—The full complement of DNA found in typical human cells, consisting of 22 autosomal chromosomes and 2 sex chromosomes (X and Y), totaling approximately 3.1 billion DNA nucleotides. The human mitochondrial DNA sequence is also considered part of the human genome.

I allele—Refers to the insertion allele of an insertion/deletion polymorphism.

identical twins—See monozygotic twins.

independent assortment—The random assortment of paternal and maternal chromosomes to the gametes, ensuring a novel genetic sequence combination in an offspring.

initiation codon—The specific three-nucleotide sequence in mRNA that is recognized as the start signal for translation. AUG, coding for the amino acid methionine, is the initiation codon.

insertion/deletion polymorphism—Genetic variation in which the alleles consist of either the presence or absence of a specific sequence of DNA, from one to several hundred nucleotides in length.

interindividual variability—Differences in phenotype or trait measurement values among individuals. Such variability can be due to experimental error, environmental factors, and genetic factors.

intermediate phenotype—A trait that contributes to a more complex trait but is influenced by fewer genetic and environmental factors; also known as a subphenotype. Multiple intermediate phenotypes will contribute to a more complex trait.

intron—A region of DNA sequence that separates two exons or coding regions within a gene.

in vitro fertilization—The process of conception (combining sperm and egg) outside of the woman's womb.

kinesiogenomics—The study of physical activity or exercise interactions with specific DNA sequence variations, such that exercise interventions may be specifically prescribed to individuals with specific genotypes with the goal of maximizing health benefits.

linkage analysis—A statistical procedure used to identify genomic regions correlated with a phenotype in related individuals (i.e., family members).

linkage disequilibrium—The nonrandom association of distinct alleles at two or more polymorphisms.

linkage map—A graphical representation of a linkage analysis, showing the extent to which specific genomic regions in each chromosome across the genome correlate with a particular phenotype.

locus (loci)—The specific location(s) of a gene or DNA sequence on a chromosome.

LOD score—The statistical measure used to identify genomic regions correlated with a phenotype in a linkage analysis study.

mean—See average.

meiosis—The process of DNA replication, chromosomal recombination, and cell division that occurs in germ cells leading to the production of gametes, each of which contains half of the genetic information needed for organism development.

Mendelian disease or trait—A phenotype that is determined by a single gene, especially as it relates to disease gene mutations; also known as a single-gene disease or trait.

Mendelian inheritance pattern—The offspring of two parents can be predicted to carry certain combinations of genetic information depending on the DNA sequence carried by each parent. These tenets of inheritance were first discovered by Gregor Mendel.

messenger RNA (mRNA)—A ribonucleic acid sequence complementary to a DNA sequence that is used for the coding of a protein.

microarray—A collection of microscopic DNA probes on a solid surface used to study RNA levels (gene expression) or identify genotypes for many polymorphisms. The microscopic nature of the tool allows for extremely large numbers of probes to be studied in a single experiment.

microsatellite repeat—See repeat polymorphism.

migration—An evolutionary force resulting in a change in DNA sequence over many generations due to the movement of a subgroup of individuals from a larger population. Offspring of the migrated subgroup will carry the genetic variation specific to that subgroup, which may differ from that of the original population.

missense polymorphism—DNA sequence variation in the coding region of a gene that carries at least two different alleles, such that each allele results in a different amino acid for the protein sequence.

mitosis—The process of DNA replication and cell division performed in somatic cells that results in the formation of two daughter cells.

monozygotic twins—Two individuals born together after a single fertilized egg splits early in the process of development, such that two independent zygotes develop with the same DNA sequence. Monozygotic twins are distinct from dizygotic twins.

mutagen—A physical or chemical agent that disrupts the genetic information of an individual, thus resulting in a mutation of the genetic information (e.g., novel DNA sequence mutation).

mutation—An evolutionary force resulting in a change in DNA sequence due to a novel change in the DNA that occurs as a result of a chemical or environmental mutagen or an error in DNA replication. Mutation also refers to a genetic sequence variation with two or more alleles present at a particular DNA sequence position in different individuals. A mutation differs from a polymorphism in that the rare allele is observed in less than 1% of the population.

natural selection—An evolutionary force resulting in a change in DNA sequence over many generations due to the enhanced survival and reproduction capacity of individuals with a distinct phenotype advantage based on a specific genetic variation.

nonsense polymorphism—DNA sequence variation in the coding region of a gene that carries at least two different alleles, such that one allele codes for one of the three termination, or stop, codons.

nucleotide—A nucleic acid molecule, also called a base, that acts as a building block for DNA and RNA sequences.

nutrigenomics—The study of nutrient interactions with specific DNA sequence variations, such that dietary factors may be specifically prescribed to individual genotypes with the goal of reducing disease risk and improving health.

penetrance—The extent to which a particular DNA sequence variation will affect a phenotype. A highly penetrant disease-causing allele will result in disease nearly every time it is carried in an individual, while an allele with low penetrance is less likely to result in disease when carried by an individual.

personalized medicine—The use of individual genetic information in the management of health care in individuals.

pharmacogenomics—The study of prescription drug interactions with specific DNA sequence variations, such that prescriptions and dosages may be tailored specifically to individual genotypes with the goal of improving effectiveness and reducing side effects.

phase—The presence of two or more alleles at two linked polymorphisms that are on the same chromosomes. Alleles are said to be out of phase when they are present on different chromosomes.

phenotype—Any observable or measurable trait, often determined by the combination of genetic and environmental factors; also known as a trait.

polygenic—Referring to phenotypes that are regulated by multiple genetic factors.

polymerase chain reaction (PCR)—A laboratory method used to make many copies of a specific DNA sequence.

polymorphism—A genetic sequence variation with two or more alleles present at a particular DNA sequence position in different individuals. A polymorphism differs from a mutation in that the rare allele is observed in more than 1% of the population.

population stratification—The extent to which DNA sequences differ among individuals of a group due to different geographical ancestries specific to each individual in the group.

posttranscriptional modification—The cellular processes performed on an RNA sequence after transcription and before translation, including the removal of intron sequences.

posttranslational modification—The cellular processes performed on an amino acid sequence after translation, resulting in the formation of the mature protein.

premature stop codon—A novel termination, or stop, codon produced by a nonsense allele in the coding region of a gene, which results in a truncated version of the expected protein sequence.

promoter—The region of DNA sequence directly upstream of the first exon, which contains regulatory sequences that are required for transcription of the gene.

protein—A linear molecule composed of several amino acids connected in a specific sequence; also known as a polypeptide.

PubMed—A free online search engine operated by the National Library of Medicine, which provides access to the citations of the Medline database from 1966 to the present. The Medline database contains literature from a wide range of health-related fields.

recessive—Referring to that allele of a polymorphism that does not determine the phenotype when in the presence of the other, dominant allele, despite the presence of both alleles in the cell. Two copies of a recessive disease allele are required for disease to develop. Recessive is the opposite of dominant.

recombination block—A large region of DNA sequence that is exchanged as one piece in the process of crossover during meiosis. Recombination blocks are distinguished from haplotype blocks in that they are family-level recombination events rather than population-level recombination events.

recombination hotspot—The boundaries of recombination or haplotype blocks that correspond to crossover events, where DNA sequence regions are exchanged between homologous chromosomes.

repeat polymorphism—DNA sequence variant that consists of multiple repeats of a specific sequence of nucleotides (e.g., **CAG**CAGCAGCAG . . .), the alleles of which are distinguished as the number of repeats in a particular sequence.

restriction enzyme—An enzyme that cleaves a DNA molecule at a sequence of nucleotides specific to that enzyme. A number of such enzymes exist, each with a specific recognition sequence that signals the enzyme to cleave the DNA molecule at that location.

restriction fragment length polymorphism (RFLP)—A variation in DNA sequence that can be deciphered using a combination of the polymerase chain reaction and specific restriction enzymes.

ribonucleic acid (RNA)—A combination of nucleic acids in series, typically single stranded. RNA consists of the nucleotides adenine, guanine, cytosine, and uracil.

ribosome—The cellular machine used in the process of translation, whereby an mRNA molecule is read and used to manufacture an amino acid sequence.

RNA polymerase—The cellular enzyme that transcribes DNA sequence into a complementary RNA sequence.

selection—See natural selection.

sex cells—The sperm in males and the ova, or egg cells, in females, which each contain half the required genetic material for an individual, and which combine to form a new offspring with the full complement of genetic material.

sex chromosomes—The two chromosomes, X and Y, that carry the genes important for sex determination. Females carry two copies of the X chromosome (XX), whereas males carry one copy of each sex chromosome (XY).

silent polymorphism—DNA sequence variation in the coding region of a gene that carries at least two different alleles, such that each allele codes for the same amino acid in the protein sequence.

single gene disease—See Mendelian disease or trait.

single-nucleotide polymorphism (SNP)—A DNA sequence variation that consists of two different alleles, each of which is only one nucleotide in length. Pronounced "snip."

somatic cell—The cells important to the formation of the body, but not responsible for the formation of new offspring. Somatic cells are distinct from germ cells.

somatic cell gene therapy—The use of gene therapy on somatic cells, which have no possibility of transmitting the therapeutic gene product to future generations.

splicing—The cellular process whereby intron sequences are removed from an RNA sequence and the exons are joined together to form a continuous strand of coding mRNA sequence in preparation for translation of the protein.

start codon—See initiation codon.

stem cells—Undifferentiated cells that have the potential to develop or differentiate into multiple cell types.

stop codon—See termination codon.

subphenotype—See intermediate phenotype.

tag SNP—A single-nucleotide polymorphism (SNP) within a haplotype block in strong linkage disequilibrium with nearby polymorphisms, such that identifying the alleles within that particular SNP allows prediction of the alleles carried at the nearby polymorphisms. Tag SNPs are used in the identification of haplotypes.

termination codon—A three-nucleotide codon sequence that signals the ribosome to stop the translation of protein from an mRNA sequence. Three such termination codons exist in the genetic code: UAA, UAG, and UGA.

terminator region—The region of DNA sequence directly downstream of the last exon, which contains regulatory sequences that are required for ending transcription of the gene.

thymine (T)—A nucleotide base found in DNA. Thymine binds with adenine (A) in the DNA sequence. Thymine is replaced by uracil (U) in the RNA sequence.

trait—Any observable or measurable characteristic or phenotype, often determined by the combination of genetic and environmental factors.

transcription—The process whereby the RNA polymerase enzyme manufactures a complementary RNA strand from a DNA sequence in the cell nucleus; also known as gene expression.

transcription factor proteins—Proteins that bind to the promoter region of a gene in the DNA sequence to promote or inhibit the process of transcription.

translation—The process whereby the ribosome manufactures an amino acid sequence complementary to an mRNA sequence in the cell cytoplasm.

untranslated region—RNA sequence segments that exist before and after the coding region of the sequence that are not translated into an amino acid sequence. Such regions typically contain signal sequences important for mRNA regulation.

uracil (U)—A nucleotide base found in RNA, in place of thymine (T) found in DNA. Uracil shares complementary binding with adenine (A).

variation—Refers to the extent of the difference observed for trait values in a set of individuals; the distance between the highest and lowest values.

Bibliography

Abbott, A. With your genes? Take one of these, three times a day. *Nature* 425: 760-762, 2003.

Abbott, A. All pain, no gain? *Nature* 433: 188-189, 2005.

Alberts, B., D. Bray, J. Lewis, M. Raff, K. Roberts, J.D. Watson. *Molecular biology of the cell,* 3rd edition. London: Garland, 1994.

Altshuler, D., A.G. Clark. Harvesting medical information from the human family tree. *Science* 307: 1052-1053, 2005.

Baldwin, K. Research in the exercise sciences: Where do we go from here? *J. Appl. Physiol.* 88: 332-336, 2000.

Behar, M. Will genetics destroy sports? *Discover* 25 (7): 41-45, 2004.

Bell, J. The double helix in clinical practice. *Nature* 421: 414-416, 2003.

Bouchard, C., P. An, T. Rice, J.S. Skinner, J.H. Wilmore, J. Gagnon, L. Perusse, A.S. Leon, D.C. Rao. Familial aggregation of $\dot{V}O_2$max response to exercise training: Results from the HERITAGE Family Study. *J. Appl. Physiol.* 87: 1003-1008, 1999.

Bouchard, C., R.M. Malina, L. Perusse. *Genetics of fitness and physical performance.* Champaign, IL: Human Kinetics, 1997.

Bouchard, C., T. Rankinen. Individual differences in response to regular physical activity. *Med. Sci. Sports Exerc.* 33: S446-S451, 2001.

Bouchard, C., T. Rankinen, Y.C. Chagnon, T. Rice, L. Perusse, J. Gagnon, I. Borecki, P. An, A.S. Leon, J.S. Skinner, J.H. Wilmore, M.A. Province, D.C. Rao. Genomic scan for maximal oxygen uptake and its response to training in the HERITAGE Family Study. *J. Appl. Physiol.* 88: 551-559, 2000.

Burke, W. Genomics as a probe for disease biology. *New Eng. J. Med.* 349: 969-974, 2003.

Chakravarti, A., P. Little. Nature, nurture and human disease. *Nature* 421: 412-414, 2003.

Clark, A.G. The role of haplotypes in candidate gene studies. *Genet. Epidemiol.* 27: 321-333, 2004.

Clayton, E.W. Ethical, legal, and social implications of genomic medicine. *New Eng. J. Med.* 349: 562-569, 2003.

Collins, F.S. What we do and don't know about "race," "ethnicity," genetics and health at the dawn of the genome era. *Nature Genet.* 36 (Suppl.): S13-S15, 2004.

Collins, F.S., M. Morgan, A. Patrinos. The Human Genome Project: lessons from large-scale biology. *Science* 300: 286-290, 2003.

Crawford, D.C., D.A. Nickerson. Definition and clinical importance of haplotypes. *Annual Rev. Med.* 56: 303-320, 2005.

Debusk, R.M., C.P. Fogarty, J.M. Ordovas, K.S. Kornman. Nutritional genomics in practice: Where do we begin? *J. Am. Diet. Assoc.* 105: 589-598, 2005.

de la Chapelle, A., A.-L. Traskelin, E. Juvonen. Truncated erythropoietin receptor causes dominantly inherited benign human erythrocytosis. *Proc. Nat. Acad. Sci. USA* 90: 4495-4499, 1993.

Dennis, C. Altered states. *Nature* 421: 686-688, 2003.

Dennis, C. The rough guide to the genome. *Nature* 425: 758-759, 2003.

Dennis, C. Rugby team converts to give gene tests a try. *Nature* 434: 260, 2005.

Gelehrter, T.D., F.S. Collins, D. Ginsburg. *Principles of medical genetics,* 2nd edition. Baltimore: Williams & Wilkins, 1998.

Gibney, M.J., E.R. Gibney. Diet, genes and disease: Implications for nutrition policy. *Proc. Nutr. Soc.* 63: 491-500, 2004.

Greely, H.T. Banning genetic discrimination. *New Eng. J. Med.* 353: 865-867, 2005.

Grigorenko, E.L. The inherent complexities of gene-environment interactions. *J. Gerontol.* 60B (Spec. Issue I): 53-64, 2005.

Guttmacher, A.E., F.S. Collins. Genomic medicine—A primer. *New Eng. J. Med.* 347: 1512-1527, 2002.

Guttmacher, A.E., F.S. Collins. Welcome to the genomic era. *New Eng. J. Med.* 349: 996-998, 2003.

Haga, S.B., M.J. Khoury, W. Burke. Genomic profiling to promote a healthy lifestyle: Not ready for prime time. *Nature Genet.* 34: 347-350, 2003.

Haga, S., J.C. Venter. FDA races in wrong direction. *Science* 301: 466, 2003.

Holden, C. Race and medicine. *Science* 302: 594-596, 2003.

Hubal, M.J., H. Gordish-Dressman, P.D. Thompson, T.B. Price, E.P. Hoffman, T.J. Angelopoulos, P.M. Gordon, N.M. Moyna, L.S. Pescatello, P.S. Visich, R.F. Zoeller, R.L. Seip, P.M. Clarkson. Variability in muscle size and strength gain after unilateral resistance training. *Med. Sci. Sports Exerc.* 37: 964-972, 2005.

Hunter, D.J. Gene-environment interactions in human diseases. *Nat. Rev. Genet.* 6: 287-298, 2005.

International HapMap Consortium. A haplotype map of the human genome. *Nature* 437: 1299-1320, 2005.

International Human Genome Sequencing Consortium, The initial sequencing and analysis of the human genome. *Nature* 409: 860-921, 2001.

Jorde, L.B., S.P. Wooding. Genetic variation, classification and "race." *Nature Genet.* 36 (Suppl.): S28-S33, 2004.

Kaplan, J.B., T. Bennett. Use of race and ethnicity in biomedical publication. *J. Am. Med. Assoc.* 289: 2709-2716, 2003.

Little, P.F.R. Structure and function of the human genome. *Genome Res.* 15: 1759-1766, 2005.

MacArthur, D.G., K.N. North. Genes and human elite athletic performance. *Hum. Genet.* 116: 331-339, 2005.

Mayeux, R. Mapping the new frontier: Complex genetic disorders. *J. Clin. Invest.* 115: 1404-1407, 2005.

McKusick, V.A. The anatomy of the human genome: A neo-vesalian basis for medicine in the 21st century. *J. Am. Med. Assoc.* 286: 2289-2295, 2001.

McPherron A.C., S-J Lee. Double muscling in cattle due to mutations in the myostatin gene. *Proc. Nat. Acad. Sci. USA* 94: 12457-12461, 1997.

Meirhaeghe, A., N. Helbecque, D. Cottel, P. Amouyel. B2-adrenoceptor gene polymorphism, body weight, and physical activity. *Lancet* 353: 896, 1999.

Merikangas, K.R., N. Risch. Genomic priorities and public health. *Science* 302: 599-601, 2003.

Miah, A. *Genetically modified athletes.* Oxon, UK: Routledge, 2004.

Morrison, A., R. Levy. Toward individualized pharmaceutical care of East Asians: The value of genetic testing for polymorphisms in drug-metabolizing genes. *Pharmacogenomics* 5: 673-689, 2004.

Muller, M., S. Kersten. Nutrigenomics: Goals and strategies. *Nat. Rev. Genet.* 4: 315-322, 2003.

Newton-Cheh, C., J.N. Hirschhorn. Genetic association studies of complex traits: Design and analysis issues. *Mutation Res.* 573: 54-69, 2005.

Niemi, A.K., K. Majamaa. Mitochondrial DNA and ACTN3 genotypes in Finnish elite endurance and sprint athletes. *Eur. J. Hum. Genet.* 13: 965-969, 2005.

Page, G.P., V. George, R.C. Go, P.Z. Page, D.B. Allison. "Are we there yet?" Deciding when one has demonstrated specific genetic causation in complex diseases and quantitative traits. *Am. J. Hum. Genet.* 73: 711-719, 2003.

Pollard, T.D. The future of biomedical research: From the inventory of genes to understanding physiology and the molecular basis of disease. *J. Am. Med. Assoc.* 287: 1725-1727, 2002.

Pray, L.A. Epigenetics: Genome, meet your environment. *Scientist,* July 5, 2004.

Pray, L.A. Dieting for the genome generation. *Scientist,* Jan 17, 2005.

Rankinen, T., M.S. Bray, J.M. Hagberg, L. Perusse, S.M. Roth, B. Wolfarth, C. Bouchard. The human gene map for performance and health-related fitness phenotypes: The 2005 update. *Med. Sci. Sports Exerc.* 38(11): 1863-1888, 2006.

Rankinen, T., A. Zuberi, Y.C. Chagnon, S.J. Weisnagel, G. Argyropoulos, B. Walts, L. Perusse, C. Bouchard. The human obesity gene map: The 2005 update. *Obesity* 14(4): 529-644, 2006.

Robertson, K.D. DNA methylation and human disease. *Nat. Rev. Genet.* 6: 597-610, 2005.

Roche, P.A., G.J. Annas. Protecting genetic privacy. *Nat. Rev. Genet.* 2: 392-396, 2001.

Rosenberg, N.A., J.K. Pritchard, J.L. Weber, H.M. Cann, K.K. Kidd, L.A. Zhivotovsky, M.W. Feldman. Genetic structure of human populations. *Science* 298: 2381-2385, 2002.

Sadee, W., Z. Dai. Pharmacogenetics/genomics and personalized medicine. *Hum. Mol. Genet.* 14: R207-R214, 2005.

Schuelke, M., K.R. Wagner, S. L.E, C. Hubner, T. Riebel, W. Komen, T. Braun, J.F. Tobin, and S-J Lee. Myostatin mutation associated with gross muscle hypertrophy in a child. *N. Eng. J. Med.* 350: 2682-2688, 2004.

Schuit, A.J., E.J.M. Feskens, L.J. Launer, and D. Kromhout. Physical activity and cognitive decline, the role of the apolipoprotein ϵ4 allele. *Med. Sci. Sports Exerc.* 33: 772-777, 2001.

Scriver, C.R., A.L. Beaudet, W.S. Sly, D. Valle (Editors); B. Childs, K.W. Kinzler, B. Vogelstein (Assoc. Editors). *The metabolic and molecular bases of inherited disease,* 8th edition. New York: McGraw-Hill, 2001.

Service, R.F. Going from genome to pill. *Science* 308: 1858-1860, 2005.

Subramanian, G., M.D. Adams, J.C. Venter, S. Broder. Implications of the human genome for understanding human biology and medicine. *J. Am. Med. Assoc.* 286: 2296-2307, 2001.

Surbone, A. Genetic medicine: The balance between science and morality. *Annals Oncol.* 15 (Suppl. 1): i60-i64, 2004.

Tabor, H.K., N.J. Risch, R.M. Myers. Candidate-gene approaches for studying complex genetic traits: Practical considerations. *Nat. Rev. Genet.* 3: 391-397, 2002.

Valle, D. Genetics, individuality, and medicine in the 21st century. *Am. J. Hum. Genet.* 74: 374-381, 2004.

Venter, J.C., et al. The sequence of the human genome. *Science* 291: 1304-1351, 2001.

Weinshilboum, R. Inheritance and drug response. *New Eng. J. Med.* 348: 529-537, 2003.

Willett, W. Balancing life-style and genomics research for disease prevention. *Science* 296: 695-703, 2002.

Wolfarth, B., M.S. Bray, J.M. Hagberg, L. Perusse, R. Rauramaa, M.A. Rivera, S.M. Roth, T. Rankinen, and C. Bouchard. The human gene map for performance and health-related fitness phenotypes: The 2004 update. *Med. Sci. Sports Exerc.* 37: 881-903, 2005.

Wong, A.H.C., I.I. Gottesman, A. Petronis. Phenotypic differences in genetically identical organisms: The epigenetic perspective. *Hum. Mol. Genet.* 14: R11-R18, 2005.

Yang, N., D.G. MacArthur, J.P. Gulbin, A.G. Hahn, A.H. Beggs, S. Easteal, and K.N. North. ACTN3 genotype is associated with human elite athletic performance. *Am. J. Hum. Genet.* 73: 627-631, 2003.

Index

Note: The italicized *f* and *t* following page numbers refer to figures and tables respectively.

About the Author

Photo courtesy of the University of Maryland Photo Services.

Stephen M. Roth, PhD, is an assistant professor in the department of kinesiology at the University of Maryland, where he teaches courses in genetics and exercise science and health. He has extensive training in both exercise science and human genetics, and he directs the Functional Genomics Laboratory, where he maintains a research program dedicated to understanding the role of genetics in the contexts of aging, health, and exercise.

Dr. Roth has more than 45 peer-reviewed publications in scholarly journals, many of which are in the areas of genetics, health, and exercise. In 2005, the American College of Sports Medicine presented him with a New Investigator Award recognizing his work in genetics and exercise physiology. He also has an award from the National Institutes of Health to further his training in genetic epidemiology, extending his existing training in exercise physiology and human genetics.

Dr. Roth is a member of the American College of Sports Medicine, the American Physiological Society, and the American Society of Human Genetics. He is an assistant editor for *Exercise and Sport Sciences Reviews.*

UWIC LEARNING CENTRE
LIBRARY DIVISION-LLANDAFF
WESTERN AVENUE
CARDIFF
CF5 2YB

You'll find other outstanding exercise science resources at

www.HumanKinetics.com

In the U.S. call

1-800-747-4457

Australia..08 8372 0999
Canada..1-800-465-7301
Europe..+44 (0) 113 255 5665
New Zealand..0064 9 448 1207

HUMAN KINETICS
The Information Leader in Physical Activity
P.O. Box 5076 • Champaign, IL 61825-5076 USA